1980

Elementary
Number Theory

A Series of Books in Mathematics

Editors:
R. A. Rosenbaum
G. Philip Johnson

Elementary
Number Theory

UNDERWOOD DUDLEY

DePauw University

W. H. FREEMAN AND COMPANY

San Francisco

Printed in the United States of America

Standard Book Number: 7167 0438-2

Library of Congress Catalog Card Number: 71-80080

9 8 7 6 5 4 3

Contents

Preface

Today, when a course in number theory is offered at all, it is usually taken only by mathematics majors late in their undergraduate careers. There is no reason why number theory should not be more widely taught, since it is, I think, manifestly more valuable to a prospective mathematics teacher than a course in, say, differential equations. This text has been designed for a one-semester or one-quarter course in number theory with minimal prerequisites. The reader is not required to know any mathematics except for the properties of the real numbers and elementary algebra. (A small amount of analysis is helpful, but not necessary, in Sections 21 and 22.) Nevertheless, because the average student does not find number theory easy, I have written detailed proofs and included many numerical examples. The purpose of the examples is to illustrate the theorems and also to try to show the fascination of playing with numbers, which is at the bottom of many of the theorems.

I have included here at least an introduction to most of the topics of elementary number theory. In Sections 1 through 5 the fundamental properties of the integers and congruences are developed, and in Section 6 proofs of Fermat's and Wilson's theorems are given. The number-theoretic functions d, σ, and ϕ are introduced in Sections 7 to 9. Sections 10 to 12 culminate in the quadratic reciprocity theorem. There follow three more or less independent blocks of material: on the representation of numbers (Sections 13 to 15), diophantine equations (16 to 20), and primes (21 to 22). Because I think that problems and exercises are especially important and interesting in number theory, Section 23 consists of 105 miscellaneous problems. They are arranged, very roughly, in increasing order of difficulty and without regard to topic.

For the typical class of inexperienced students, there is more than enough material for one semester. Any of the last three blocks of material mentioned above could be omitted. For that matter, Sections 10 to 12 (which contain the most difficult material in the book) could be left out, since the only fact proved there and used later is that -1 is a quadratic residue or nonresidue of an odd prime p according as $p \equiv 1$ or 3 (mod 4).

Proofs of theorems are often neither the shortest nor the most elegant known, but are what I thought the most natural. For example, the adaption of Tchebyshev's theorem on the bounds of $\pi(x)$ in section 22 is quite long, but it has the advantage of introducing functions and techniques used elsewhere in number theory. Besides, when the student sees the more elegant proof of the theorem, he will be all the more impressed.

In keeping with the idea that the only way to learn mathematics is to do it, I have included more than a thousand exercises and problems. The exercises interrupt the text (as well as some proofs), and the problems appear at the end of each section. In fact, exercises and problems appear in three of the four appendices. The exercises can be used in several ways: the student may do them as he reads the material for the first time, he may return to them later to check on his understanding of material already gone over, or the instructor may include them in his exposition.

Some of the exercises and problems are computational and some classical, but some are more or less original, and a few, I think, are startling. Because there are often many methods of attacking a problem in number theory, and because the most fruitful line may not be obvious, hints are given for many problems. These hints appear in a section at the back of the book (some hints are almost complete solutions). A solution that does not use a hint is of course more praiseworthy than one that does, but the amount of inspiration necessary to solve some problems is too large to be had to order. Number theory problems can be hard: I. A. Barnett has written (in the November 1966 *American Mathematical Monthly*), "To discover mathematical talent, there is no better course in elementary mathematics than number theory. Any student who can work the exercises in a modern text in number theory should be encouraged to pursue a mathematical career." Answers are provided for some problems and exercises in separate sections at the back of the book. Although there are more problems than a student could solve in one semester, he is encouraged to treat them, along with their hints, as part of the text— to be read with almost equal attention. Often they may be found more interesting than the material on which they are based.

The book contains four appendices. The first two (Appendix A, Proof by Induction, and Appendix B, Notations) should be read by the student when necessary. Appendix C (Quadratic Congruences with Composite Moduli), is included to complete the study of solutions of a quadratic congruence; it could come after section 11. Appendix D consists of three tables that are included both for their inherent fascination and for use in solving numerical problems. The first makes it easy to factor any positive integer less than 10,000, the second gives the first 447 squares, and the third consists of part of a factor table.

There are without doubt errors in the problems and in the text. I hope that when the reader finds one, he will feel pleasure at his acuteness instead of annoyance with the author. Corrections will be welcomed.

February 1969   UNDERWOOD DUDLEY

Integers

A large part of number theory is devoted to studying the properties of the integers—that is, the numbers . . . , $-2, -1, 0, 1, 2, \ldots$. Usually the integers are used merely to convey information (3 apples, \$32, $17x^2 + 9$), with no consideration of their properties. When counting apples, dollars, or x^2's, it is immaterial how many divisors 3 has, whether 32 is prime or not, or that 17 can be written as the sum of the squares of two integers. But the integers are so basic a part of mathematics that they have been thought worthy of study for their own sake. The same situation arises elsewhere: the number theorist is comparable to the linguist, who studies words and their properties, independent of their meaning. Another justification for studying the integers is esthetic: many beautiful results can be proved.

In this section, and until further notice, lower-case *italic* letters will invariably denote integers. We will take as known and use freely the usual properties of addition, subtraction, multiplication, division, and order for the integers. We also use in this section an important property of the integers—a property that you may not be consciously aware of because it is not stated explicitly as often as, say, the associative property of multiplication. It is the *least-integer principle*: a nonempty set of integers that is bounded below contains a smallest element. It is also true that a nonempty set of integers that is bounded above contains a largest element.

We will say that *a divides b* (written $a \mid b$) if and only if there is an integer d such that $ad = b$. For example, $2 \mid 6$, $12 \mid 60$, $17 \mid 17$, $-5 \mid 50$, and $8 \mid -24$. If a does not divide b, we will write $a \nmid b$. For example, $4 \nmid 2$ and $3 \nmid 4$.

Exercise 1. Which integers divide zero?

Exercise 2. Show that if $a \mid b$ and $b \mid c$, then $a \mid c$.

As a sample of the sort of properties that division has, we prove

LEMMA 1. *If $d \mid a$ and $d \mid b$, then $d \mid (a + b)$.*

PROOF. From the definition of divides, we know that there are integers q and r such that

$$dq = a \quad \text{and} \quad dr = b.$$

Thus

$$a + b = d(q + r),$$

so from the definition again, $d \mid (a + b)$.

LEMMA 2. *If $d \mid a$ then $d \mid ca$, for any integer c.*

LEMMA 3. *If $d \mid a_1, d \mid a_2, \ldots, d \mid a_n$, then $d \mid (c_1 a_1 + c_2 a_2 + \cdots + c_n a_n)$ for any integers $c_1, c_2, \ldots, c_n$.*

The proofs are easy.

Exercise 3. Prove Lemmas 2 and 3.

As an application of Lemma 3, we see that if d divides all the terms on one side of an equation, then it divides the other side of the equation too. Thus if $a + b = c$ and $d \mid a$ and $d \mid c$, then $d \mid b$. Or, if

$$3x + 81y + 6z + 363 = w,$$

then $3 \mid w$, because 3 divides all the terms on the left-hand side of the equation. (Remember that all small *italic* letters, including x, y, z, and w, denote integers unless otherwise indicated.) Similarly, if

$$3x^2 + 15xy + 5y^2 = 0,$$

then $3 \mid 5y^2$ and $5 \mid 3x^2$.

The rest of this section will be devoted to the greatest common divisor and its properties, which we will use constantly later. We say that d is the *greatest common divisor* of a and b (written $d = (a, b)$) if and only if

(i) $d \mid a$ and $d \mid b$, and

(ii) if $c \mid a$ and $c \mid b$, then $c \leq d$.

Condition (i) says that d is a common divisor of a and b, and (ii) says that it is the greatest such divisor. Note that if a and b are not both zero, then the set of common divisors of a and b is a set of integers that is bounded above by the largest of a, b, $-a$, and $-b$. Hence, from the well-ordering principle for the integers, the set has a largest element, so the greatest common divisor of a and b exists and is unique. Note that $(0, 0)$ is not defined, and that if (a, b) is defined, then it is positive. In fact, $(a, b) \geq 1$ because $1 \mid a$ and $1 \mid b$ for all a and b.

Exercise 4. What are $(4, 14)$, $(5, 15)$, and $(6, 16)$?

Exercise 5. What is $(n, 1)$, where n is any positive integer? What is $(n, 0)$?

Exercise 6. If d is a positive integer, what is (d, nd)?

As an exercise in applying the definition of greatest common divisor, we will prove the following theorem, which we will use often later:

THEOREM 1   *If $(a, b) = d$, then $(a/d, b/d) = 1$.*

PROOF. Suppose that $c = (a/d, b/d)$. We want to show that $c = 1$. We will do this by showing that $c \leq 1$ and $c \geq 1$. The latter inequality follows from the fact that c is the greatest common divisor of two integers, and as we have noted, every greatest common divisor is greater than or equal to 1. To show that $c \leq 1$, we use the facts that $c \mid (a/d)$ and $c \mid (b/d)$. We then know that there are integers q and r such that $cq = a/d$ and $cr = b/d$. That is,

$$(cd)q = a \quad \text{and} \quad (cd)r = b.$$

These equations show that cd is a common divisor of a and b. It is thus no greater than the *greatest* common divisor of a and b, and this is d. Thus $cd \leq d$. Since d is positive, this gives $c \leq 1$. Hence $c = 1$, as was to be proved.

If $(a, b) = 1$, then we will say that a and b are *relatively prime*, for a reason that will become clear in the section on unique factorization.

When a and b are small, it is often possible to see what (a, b) is by inspection. When a and b are large, this is no longer possible. (What is $(31415926, 5358979)$?) We will now develop an efficient process for finding greatest common divisors: the Euclidean algorithm. It is also useful in proving theorems we will need later.

THEOREM 2 (*the division algorithm*). *Given positive integers a and b, $b \neq 0$, there exist unique integers q and r, with $0 \leq r < b$ such that*

$$a = bq + r.$$

PROOF. If we write $a = bq + r$ as

$$\frac{a}{b} = q + \frac{r}{b},$$

then we see that the theorem is merely a statement of what we do when we divide a by b: we get a quotient of q and a remainder of r. We will make this slightly more formal. Consider the set S of integers $a - bt$, $t = 0, \pm 1, \pm 2, \ldots$. Because S contains nonnegative elements (for example, a and $a + b$), we know from the least-integer principle for the integers that S contains a smallest nonnegative number. Call it r, and let q equal the corresponding value of t. Then $a - bq = r$, and $r \geq 0$. To complete the proof, we have to show that $r < b$. Suppose not. Then $r = b + r_1$ with $r_1 \geq 0$. Hence

$$r_1 = r - b = a - bq - b = a - b(q + 1).$$

This says that r_1 is in the set S. But

$$0 \leq r_1 = r - b < r,$$

and this is impossible because r is the *smallest* nonnegative number in the set S.

This construction gives us q and r, and it remains to show that they are unique. Suppose that we have found q, r and q_1, r_1 such that

$$a = bq + r = bq_1 + r_1$$

with $0 \leq r < b$ and $0 \leq r_1 < b$. Subtracting, we have

(1) $$0 = b(q - q_1) + (r - r_1).$$

Since b divides the left-hand side of this equation and the first term on the right-hand side, it divides the other term:

$$b \mid (r - r_1).$$

But since $0 \leq r < b$ and $0 \leq r_1 < b$, we have

$$-b < r - r_1 < b.$$

The only multiple of b between $-b$ and b is zero. Hence $r - r_1 = 0$, and it follows from (1) that $q - q_1 = 0$ too. Hence the numbers q and r in the theorem are unique.

Although the theorem was stated only for positive integers a and b, because it is most often applied for positive integers, nowhere in the proof did we need a to be positive. Moreover, if b is negative, the theorem is true if $0 \leq r < b$ is replaced with $0 \leq r < -b$; you are invited to reread the proof and verify that this is so.

Exercise 7. What are q and r if $a = 75$ and $b = 24$? If $a = 75$ and $b = 25$?

Theorem 2, combined with the next lemma, will give the Euclidean algorithm.

LEMMA 4. *If $a = bq + r$, then $(a, b) = (b, r)$.*

PROOF. Let $d = (a, b)$. We know that since $d \mid a$ and $d \mid b$, it follows from $a = bq + r$ that $d \mid r$. Thus d is a common divisor of b and r. Suppose that c is any common divisor of b and r. We know that $c \mid b$ and $c \mid r$, and it follows from $a = bq + r$ that $c \mid a$. Thus c is a common divisor of a and b, and hence $c \le d$. Both parts of the definition of greatest common divisor are satisfied, and we have that $d = (b, r)$.

Exercise 8. Verify that the lemma is true when $a = 16$ and $b = 6$, and $q = 2$.

According to Lemma 4, if we apply the division algorithm to a and b, we get $(a, b) = (b, r)$; the same greatest common divisor, but on the right there is one smaller number inside the parentheses. We can continue, and apply the division algorithm to b and r to get still smaller numbers with the same greatest common divisor. If we apply the division algorithm enough times, the numbers will eventually get so small that we can recognize the greatest common divisor by inspection. For example, let us calculate $(5767, 4453)$. Applying the division algorithm, we have

$$5767 = 4453 \cdot 1 + 1314.$$

From Lemma 4, we know that $(5767, 4453) = (4453, 1314)$. Unless you are very good at inspection, these integers are still too large for you to see their greatest common divisor. We must divide again:

$$4453 = 1314 \cdot 3 + 511.$$

Now we know that $(5767, 4453) = (1314, 511)$. We continue dividing:

$$
\begin{aligned}
1314 &= 511 \cdot 2 + 292, \\
511 &= 292 \cdot 1 + 219, \\
292 &= 219 \cdot 1 + 73, \\
219 &= 73 \cdot 3.
\end{aligned}
$$

The sequence of remainders has reached zero (as it must, since a decreasing sequence of nonnegative integers cannot go on forever), and from Lemma 4 we know that

$$(5767, 4453) = (4453, 1314) = \cdots = (219, 73) = (73, 0) = 73.$$

The formal statement of the process just carried out for a special case is

THEOREM 3 (the Euclidean algorithm). *If a and b are positive integers, $b \ne 0$, and*

$$a = bq + r, \qquad 0 \leq r < b,$$
$$b = rq_1 + r_1, \qquad 0 \leq r_1 < r,$$
$$r = r_1q_2 + r_2, \qquad 0 \leq r_2 < r_1,$$
$$\cdots \qquad\qquad \cdots$$
$$r_k = r_{k+1}q_{k+2} + r_{k+2}, \qquad 0 \leq r_{k+2} < r_{k+1},$$

then for k large enough, say k = t, we have

$$r_{t-1} = r_t q_{t+1},$$

and $(a, b) = r_t$.

PROOF. The sequence of nonnegative integers

$$b > r > r_1 > r_2 > \cdots$$

must come to an end. Eventually, one of the remainders will be zero. Suppose that it is r_{t+1}. Then $r_{t-1} = r_t q_{t+1}$. From Lemma 4 applied over and over,

$$(a, b) = (b, r) = (r, r_1) = (r_1, r_2) = \cdots = (r_{t-1}, r_t) = r_t.$$

If either a or b is negative, we can use the fact that $(a, b) = (-a, b) = (a, -b) = (-a, -b)$.

Exercise 9. Calculate $(299, 247)$ and $(578, 442)$.

A consequence of the Euclidean algorithm that will be used many times later is

THEOREM 4. *If $(a, b) = d$, then there are integers x and y such that $ax + by = d$.*

PROOF. The idea is to work the Euclidean algorithm backwards. Take, for example, the computation of $(5767, 4453) = 73$. The next-to-last line in the algorithm gives

$$73 = 292 - 219.$$

Now we use the line before that to express 73 as a combination of 511's and 292's:

$$73 = 292 - (511 - 292) = 2 \cdot 292 - 511.$$

We eliminate the 292 by using the next line before:

$$73 = 2(1314 - 511 \cdot 2) - 511 = 2 \cdot 1314 - 5 \cdot 511.$$

And so on:

$$73 = 2 \cdot 1314 - 5(4453 - 1314 \cdot 3) = 17 \cdot 1314 - 5 \cdot 4453.$$

Finally, we can express 1314 in terms of 4453 and 5767 to get the representation we want:

$$73 = 17(5767 - 4453) - 5 \cdot 4453 = 17 \cdot 5767 - 22 \cdot 4453.$$

In general, we have

$$d = (a, b) = r_t = r_{t-2} - r_{t-1}q_t;$$

this expresses d as a combination of r_{t-1} and r_{t-2} with integer coefficients. From the next preceding line of the algorithm, namely

$$r_{t-3} = r_{t-2}q_{t-1} + r_{t-1},$$

we get

$$d = r_{t-2} - (r_{t-3} - r_{t-2}q_{t-1})q_t,$$

which gives d as a combination of r_{t-2} and r_{t-3} with integer coefficients:

$$d = (q_{t-1}q_t + 1)r_{t-2} - q_t r_{t-3}.$$

We can then eliminate r_{t-2} by using

$$r_{t-4} = r_{t-3}q_{t-2} + r_{t-2},$$

and this gives

$$d = (\text{integer})r_{t-3} + (\text{integer})r_{t-4}.$$

If we continue we will eventually get integers x and y such that

$$d = ax + by.$$

Exercise 10. Find a solution of $299x + 247y = 13$.

Theorem 4 has many applications. We now give two that will be used later.

THEOREM 5. *If $d \mid ab$ and $(d, a) = 1$, then $d \mid b$.*

PROOF. Since d and a are relatively prime, we know from Theorem 4 that there are integers x and y such that

$$dx + ay = 1.$$

Multiplying this by b we have

$$d(bx) + (ab)y = b.$$

The first term on the left-hand side of the above equation can of course be divided by d, and since $d \mid ab$, then d also divides the second. Hence d divides the right-hand side too, and this is what we wanted to prove.

Note that if d and a are not relatively prime in Theorem 5, then the conclusion is false. For example, $6 \mid 8 \cdot 9$, but $6 \nmid 8$ and $6 \nmid 9$.

THEOREM 6. *Let $(a, b) = d$, and suppose that $c \mid a$ and $c \mid b$. Then $c \mid d$.*

PROOF. In words, this theorem says that every common divisor of two integers is a divisor of their greatest common divisor. The proof is short: we know that there are integers x and y such that

$$ax + by = d.$$

Since c divides each term on the left-hand side of this equation, c divides the right-hand side too.

Problems

1. Calculate (a) $(314, 159)$, (b) $(3141, 1592)$, (c) $(4144, 7696)$, (d) $(10{,}001, 100{,}083)$.

2. Prove that if $a \mid b$ and $b \mid a$, then $a = b$ or $a = -b$.

3. Show that if $a \mid b$ and $a > 0$, then $(a, b) = a$.

4. Show that $((a, b), b) = (a, b)$.

5. Show that it is false that $a > b$ implies $a \nmid b$.

6. (a) Show that $(n, n + 1) = 1$ for all $n > 0$.
(b) If $n > 0$, what values can $(n, n + 2)$ take on?
(c) If $n > 0$, what values can $(n, n + k)$ take on?

7. If $N = n_1 n_2 \cdots n_k + 1$, show that $(n_i, N) = 1$ for $i = 1, 2, \ldots, k$.

8. Show that $(a, b) = 1$ and $c \mid a$ imply $(c, b) = 1$.

9. Find x and y such that

(a) $314x + 159y = 1$,
(b) $3141x + 1592y = 1$,
(c) $4144x + 7696y = 592$,
(d) $10001x + 100083y = 73$.

10. (a) Show that $(k, n + k) = 1$ if and only if $(k, n) = 1$.
(b) Is it true that $(k, n + k) = d$ if and only if $(k, n) = d$?
(c) Is it true that $(k, n + rk) = d$ for all integers r if and only if $(k, n) = d$?

11. (a) Show that $(299, 247) = 13$.
(b) Find two solutions of $299x + 247y = 13$.
(c) Find two solutions of $299x + 247y = 52$.

12. (a) If $x^2 + ax + b = 0$ has an integer root, show that it divides b.
(b) If $x^2 + ax + b = 0$ has a rational root, show that it is in fact an integer.

13. Prove: If $a \mid b$ and $c \mid d$, then $ac \mid bd$.

14. Prove: If $d \mid a$ and $d \mid b$, then $d^2 \mid ab$.

15. Prove: If $c \mid ab$ and $(c, a) = d$, then $c \mid db$.

16. Prove: If d is odd, $d \mid (a + b)$, and $d \mid (a - b)$, then $d \mid (a, b)$.

17. Disprove: If $a \nmid b$, then $(a, b) = 1$.

18. Show that $p \mid (10a - b)$ and $p \mid (10c - d)$ imply $p \mid (ad - bc)$.

19. Prove that $6 \mid (n^3 - n)$ for all $n > 0$.

20. (a) Show that if $10 \mid (3^m + 1)$ for some m, then $10 \mid (3^{m+4n} + 1)$ for all $n > 0$.
(b) For which m does $10 \mid (3^m + 1)$?

Unique Factorization

The aim of this section is to introduce the prime numbers, which are one of the main objects of study in number theory, and to prove the unique factorization theorem for positive integers, which is essential in what comes later. In this section, lower-case italic letters invariably denote *positive* integers.

A *prime* is an integer that is greater than 1 and has no positive divisors other than 1 and itself. An integer that is greater than 1 but is not prime is called *composite*. Thus 2, 3, 5, and 7 are prime, and 4, 6, 8, and 9 are composite. There are also large primes:

$$170,141,183,460,469,231,731,687,303,715,884,105,727$$

is one, and it is clear that there are arbitrarily large composite numbers. Note that we call 1 neither prime nor composite. Although it has no positive divisors other than 1 and itself, including it among the primes would make the statement of some theorems inconvenient, in particular the unique factorization theorem. We will call 1 a *unit*. Thus the set of positive integers can be divided into three classes: the primes, the composites, and a unit.

Exercise 1. How many even primes are there? How many whose last digit is 5?

Our aim is to show that each positive integer can be written as a product of primes—and, moreover, in only one way. We will not count products that differ only in the order of their factors as different factorizations. Thus we will consider each of

$$2^2 \cdot 3 \cdot 7, \qquad 2 \cdot 3 \cdot 7 \cdot 2, \qquad \text{and} \qquad 7 \cdot 3 \cdot 2 \cdot 2$$

as the same factorization of 84. The primes can thus be used to build, by multiplication, the entire system of positive integers. The first two lemmas that follow will show that every positive integer can be written as a product of primes. Later we will prove the uniqueness of the representation.

LEMMA 1. *Every integer n, $n > 1$, is divisible by a prime.*

PROOF. If n is a prime, then the lemma is proved, because n divides itself. Suppose, however, that n is composite. Then, by definition, n has a divisor other than 1 and itself. Suppose that d_1 is that divisor; then $n = d_1 n_1$ for some integer n_1, and because $d_1 \neq 1$ or n, it follows that $1 < n_1 < n$. (In fact, $n_1 \leq n/2$, but we need only the fact that it is smaller than n.) If n_1 is prime, then $n_1 \mid n$ and the lemma is proved, since we have found a prime divisor of n. But if n_1 is composite, then $n_1 = d_2 n_2$ for some integer n_2 with $1 < n_2 < n_1$. If n_2 is a prime, then we need go no further: n_2 is a prime and $n_2 \mid n$ (because $n_2 \mid n_1$ and $n_1 \mid n$). If n_2 is not prime, then it is composite (recall that n_2 is greater than 1), and $n_2 = d_3 n_3$ with $1 < n_3 < n_2$. And so on: among the numbers $n, n_1, n_2, n_3, \ldots$ a prime will eventually appear, because

$$n > n_1 > n_2 > n_3 > n_4 > \cdots$$

and each n_i is greater than 1. It is impossible for a decreasing sequence of positive integers to continue forever. The prime that eventually appears—call it n_k—divides n because

$$n_k \mid n_{k-1}, \qquad n_{k-1} \mid n_{k-2}, \qquad \ldots, \qquad n_1 \mid n$$

imply $n_k \mid n$.

Lemma 1 can also be proved, more efficiently, using the second principle of induction (see Appendix A). The lemma is true by inspection for $n = 2$. Suppose it is true for $n \leq k$. Then either $k + 1$ is prime, in which case we are done, or it is divisible by some number k_1 with $k_1 \leq k$. But from the induction assumption, k_1 is divisible by a prime, and this prime also divides $k + 1$. Again, we are done. This proof is essentially the same as the first; the principle of induction replaces the "and so on" used there.

With the aid of Lemma 1 and an argument similar to the one used in its proof, we can prove that every positive integer can be written as a product of primes in at least one way.

LEMMA 2. *Every integer n, $n > 1$, can be written as a product of primes.*

PROOF. From Lemma 1, we know that there is a prime p_1 such that $p_1 \mid n$. That is, $n = p_1 n_1$, where $1 \leq n_1 < n$. If $n_1 = 1$, then we are done: $n = p_1$ is an expression of n as a product of primes. If $n_1 > 1$, then from

Lemma 1 again, there is a prime that divides n_1. That is, $n_1 = p_2 n_2$, where p_2 is a prime and $1 \leq n_2 < n_1$. If $n_2 = 1$, again we are done: $n = p_1 p_2$ is written as a product of primes. But if $n_2 > 1$, then Lemma 1 once again says that $n_2 = p_3 n_3$, with p_3 a prime and $1 \leq n_3 < n_2$. If $n_3 = 1$, we are done. If not we continue. We will sooner or later come to one of the n_i equal to 1, because $n > n_1 > n_2 > \cdots$ and each n_i is positive; such a sequence cannot continue forever. For some k, we will have $n_k = 1$, in which case $n = p_1 p_2 \cdots p_k$ is the desired expression of n as a product of primes. Note that the same prime may occur several times in the product.

Exercise 2 (optional). Construct a proof of Lemma 2 using induction.

Exercise 3. Write prime decompositions for 72 and 480.

Before we show that each positive integer has only one prime decomposition, we will prove an old and elegant theorem:

THEOREM 1 (Euclid). *There are infinitely many primes.*

PROOF. Suppose not. Then there are only finitely many primes. Denote them by $p_1, p_2, \ldots, p_r$. Consider the integer

$$(1) \qquad n = p_1 p_2 \cdots p_r + 1.$$

From Lemma 1, we see that n is divisible by a prime, and since there are only finitely many primes, it must be one of $p_1, p_2, \ldots, p_r$. Suppose that it is p_k. Then since

$$p_k \mid n \qquad \text{and} \qquad p_k \mid p_1 p_2 \cdots p_r,$$

it divides two of the terms in (1). Consequently it divides the other term in (1); thus $p_k \mid 1$. This is nonsense: no primes divide 1 because all are greater than 1. This contradiction shows that we started with an incorrect assumption. Since there cannot be only finitely many primes, there are infinitely many.

This is a strong theorem. We can actually identify only finitely many primes—the largest prime currently known is $2^{11213} - 1$, and we by no means know all of the primes smaller than this one. (There are lists of all the primes smaller than 100,000,000, but none that go much further.) The prime $2^{11213} - 1$ is considerably larger: it has 3,376 digits. Although $2^{11213} - 1$ is a large number, there are infinitely many integers larger than it, and only finitely many smaller. Thus, although we can name only finitely many primes, we may be sure that no matter how many we discover, there is always one more that we have yet to find. Before the development of high-speed computers, the largest prime known was the comparatively puny 39-digit number displayed at the beginning of this section. Hence if you set out to find a prime larger than $2^{11213} - 1$ without the aid of machinery, you will need a great deal of time to spare—several centuries at the least.

We will make one more digression before proving the unique factorization theorem; we show how to construct a table of prime numbers.

LEMMA 3. *If n is composite, then it has a divisor d such that $1 < d \leq n^{1/2}$.*

PROOF. Since n is composite, there are integers d_1 and d_2 such that $d_1 d_2 = n$ and $1 < d_1 < n$, $1 < d_2 < n$. If d_1 and d_2 are both larger than $n^{1/2}$, then

$$n = d_1 d_2 > n^{1/2} n^{1/2} = n,$$

which is impossible. Thus, one of d_1 and d_2 must be less than or equal to $n^{1/2}$.

LEMMA 4. *If n is composite, then it has a prime divisor less than or equal to $n^{1/2}$.*

PROOF. We know from Lemma 3 that n has a divisor—call it d—such that $1 < d \leq n^{1/2}$. From Lemma 1, we know that d has a prime divisor p. Since $p \leq d \leq n^{1/2}$, the lemma is proved.

Lemma 4 provides the basis for the following ancient method for finding primes, the well-known *Sieve of Eratosthenes*. Write the numbers from 1 to N. Put a circle around 2, and cross out all larger multiples of 2. The first number neither crossed out nor circled is 3. Put a circle around it, and cross out all its multiples. The first number now neither crossed out nor circled is 5, the next prime after 3. Put a circle around it, and cross out all its multiples. And so on: continue the process until all the numbers less than or equal to $N^{1/2}$ are crossed out or circled. Then the circled numbers in the list, together with those not crossed out, are exactly the primes less than or equal to N. Before we see why this is so, let us look at an example. Take $N = 81$; here $N^{1/2} = 9$. Sieving, we have

②	1̷4̷	2̷6̷	3̷8̷	5̷0̷	6̷2̷	7̷4̷
③	1̷5̷	2̷7̷	3̷9̷	5̷1̷	6̷3̷	7̷5̷
4̷	1̷6̷	2̷8̷	4̷0̷	5̷2̷	64	7̷6̷
⑤	17	29	41	53	6̷5̷	7̷7̷
6̷	1̷8̷	3̷0̷	4̷2̷	5̷4̷	6̷6̷	7̷8̷
⑦	19	31	43	5̷5̷	67	79
8̷	2̷0̷	3̷2̷	44	5̷6̷	6̷8̷	8̷0̷
9̷	2̷1̷	3̷3̷	4̷5̷	5̷7̷	6̷9̷	8̷1̷
1̷0̷	2̷2̷	3̷4̷	46	5̷8̷	7̷0̷	
11	23	3̷5̷	47	59	71	
1̷2̷	2̷4̷	3̷6̷	4̷8̷	6̷0̷	7̷2̷	
13	2̷5̷	37	4̷9̷	61	73	

By crossing out multiples of 2, 3, 5, and 7, we have discovered all the primes smaller than 81. Similarly, to find all the primes less than 10,000 we need only cross out the multiples of the 25 primes less than 100.

To see that this method is correct, note that any number with a circle around it is a prime, because if it were composite, it would have a prime divisor smaller than itself, and hence it would have been crossed out. Furthermore, any number not crossed out is a prime. Suppose that such a number were not a prime. Then, by Lemma 4, it would have a prime divisor less than or equal to $N^{1/2}$. But we have crossed out all of the numbers that have prime divisors less than or equal to $N^{1/2}$. Hence these and all the circled numbers are primes. None of the crossed out numbers is a prime, however, because each one has a prime divisor other than itself.

In case you are tempted to list a lot of numbers and start sieving, keep in mind that it has been tried before. In the nineteenth century, an Austrian astronomer named Kulik constructed an enormous seive of all the integers up to 100,000,000. It took him 20 years, off and on. All that work was so little valued that the library to which he left his manuscript lost the part that included the integers from 12,642,600 to 22,852,800.

The following lemma gives the result that makes unique factorization possible. For the rest of this section, and until further notice, the letters p and q will be reserved for primes.

LEMMA 5. *If $p \mid ab$, then $p \mid a$ or $p \mid b$.*

PROOF. Since p is prime, its only positive divisors are 1 and p. Thus $(p, a) = p$ or $(p, a) = 1$. In the first case, $p \mid a$, and we are done. In the second case, Theorem 5 of Section 1 tells us that $p \mid b$, and again we are done.

Exercise 4. If $p \mid a_1a_2\cdots a_k$, what can you conclude?

Exercise 5. Construct a proof by induction of the result in Exercise 4.

LEMMA 6. *If $q_1, q_2, \ldots, q_n$ are primes and $p \mid q_1q_2 \cdots q_n$, then $p = q_k$ for some k.*

PROOF. From Exercise 5 we know that $p \mid q_k$ for some k. Since p and q_k are primes, $p = q_k$. (The only positive divisors of q_k are 1 and q_k, and p is not 1.)

THEOREM 2 (*The unique factorization theorem*). *Any positive integer can be written as a product of primes in one and only one way.*

PROOF. Recall that we agree to consider as identical all factorizations that differ only in the order of the factors.

We know already from Lemma 2 that any integer n, $n > 1$, can be written as a product of primes. Thus to complete the proof of the theorem, we need to show that n cannot have two such representations. That is, if

(2) $\qquad n = p_1 p_2 \cdots p_m \qquad$ and $\qquad n = q_1 q_2 \cdots q_r,$

then we must show that the same primes appear in each product, and the same number of times, though their order may be different. That is, we must show that the integers $p_1, p_2, \ldots, p_m$ are just a rearrangement of the integers $q_1, q_2, \ldots, q_r$. From (2) we see that since $p_1 \mid n$,

$$p_1 \mid q_1 q_2 \cdots q_r.$$

From Lemma 6, it follows that $p_1 = q_i$ for some i. If we divide

$$p_1 p_2 \cdots p_m = q_1 q_2 \cdots q_r$$

by the common factor, we have

(3) $\qquad p_2 p_3 \cdots p_m = q_1 q_2 \cdots q_{i-1} q_{i+1} \cdots q_r.$

Because p_2 divides the left-hand side of this equation, it also divides the right-hand side. Applying Lemma 6 again, it follows that $p_2 = q_j$ for some $j(j = 1, 2, \ldots, i - 1, i + 1, \ldots, n)$. Cancel this factor from both sides of (3), and continue the process. Eventually we will find that each q is a p. (It is impossible for us to run out of q's before all the p's are gone, because we would then have a product of primes equal to 1, which is impossible.) If we repeat the argument with the p's and q's interchanged, we see that each q is a p. Thus the numbers $p_1, p_2, \ldots, p_m$ are a rearrangement of $q_1, q_2, \ldots, q_r$, and the two factorizations differ only in the order of the factors.

The uniqueness of the prime decomposition can also be efficiently proved by induction, though the idea is no different. The theorem is true, by inspection, for $n = 2$. Suppose that it is true for $n \leq k$. Suppose that $k + 1$ has two representations:

$$k + 1 = p_1 p_2 \cdots p_m = q_1 q_2 \cdots q_r.$$

As in the last proof, $p_1 = q_i$ for some i, so

$$p_2 p_3 \cdots p_m = q_1 q_2 \cdots q_{i-1} q_{i+1} \cdots q_r.$$

But this number is less than or equal to k, and by the induction assumption, its prime decomposition is unique. Hence the integers $q_1, q_2, \ldots, q_{i-1}, q_{i+1}, \ldots, q_r$ are a rearrangement of $p_2, p_3, \ldots, p_m$ and, since $p_1 = q_i$, the proof is complete.

Because of your long experience with the positive integers (can you remember what it was like not to know what $2 + 3$ was?), you may not find the unique factorization theorem very exciting; you may even think that it is obvious and self-evident. The following example is intended to show that it is not as self-evident as you might think: we will construct a number system in which the unique factorization theorem is not true. Consider the integers 1, 5, 9, 13, 17, . . . ; that is, all integers of the form $4n + 1$, $n = 0, 1, \ldots$. We will call an element of this set *prome* if it has no divisors other than 1 and itself *in the set*. For example, 21 is prome, whereas $25 = 5 \cdot 5$ is not.

Exercise 6. Which members of the set less than 100 are not prome?

In the same way as we proved Lemmas 1 and 2, we can show that every member of the set has a prome divisor and can be written as a product of promes. (You are invited to inspect the proofs of Lemmas 1 and 2 to see if any words need to be changed.) But an example shows that the prome decomposition of an integer in the set is not always unique:

$$693 = 21 \cdot 33 = 9 \cdot 77,$$

and 9, 21, 33, and 77 are all prome.

From the unique factorization theorem it follows that each positive integer can be written in exactly one way in the form

$$n = p_1^{e_1} p_2^{e_2} \cdots p_k^{e_k},$$

where $e_i \geq 1$, $i = 1, 2, \ldots, k$, each p_i is a prime, and $p_i \neq p_j$ for $i \neq j$. We call this representation the *prime-power decomposition* of n, and whenever we write

$$n = p_1^{e_1} p_2^{e_2} \cdots p_k^{e_k},$$

it will be understood, unless specified otherwise, that all the exponents are positive and the primes are distinct. Table A gives the smallest prime that divides n for all n less than 10,000 and not divisible by 2 or 5. With the aid of this table, the prime-power decomposition for any $n \leq 10,000$ can be found readily. For example, take 8001. It is clearly not divisible by 2 or 5, and Table A gives its smallest prime factor as 3. Then $8001/3 = 2667$, and the table shows that 3 is a factor of 2667: $2667/3 = 889$. Again referring to the table, we see that $7 \mid 889$. Finally, $889/7 = 127$, which is prime. Thus

$$8001 = 3^2 \cdot 7 \cdot 127.$$

Exercise 7. What is the prime-power decomposition of 7950?

To conclude this section, we note that the prime decomposition of integers gives another way of finding greatest common divisors besides the Euclidean algorithm. For example, consider $n = 120 = 2^3 \cdot 3 \cdot 5$ and $m = 252 = 2^2 \cdot 3^2 \cdot 7$. We see that 2^2 divides m and n, but no higher power of 2 is a common divisor of m and n. Also, 3 divides m and n, and no higher power of 3 is a common divisor. Furthermore, no other prime divides both m and n. Thus $2^2 \cdot 3$ is the greatest common divisor of m and n. Given the prime-power decompositions of m and n, we can write m and n as products of the same primes by inserting primes with the exponent zero where necessary. For example,

$$120 = 2^3 \cdot 3^1 \cdot 5^1 \cdot 7^0 \quad \text{and} \quad 252 = 2^2 \cdot 3^2 \cdot 5^0 \cdot 7^1.$$

In general, we have

THEOREM 3. *If* $e_i \geq 0, f_i \geq 0, (i = 1, 2, \ldots, k),$

(4) $\qquad\qquad m = p_1^{e_1} p_2^{e_2} \cdots p_k^{e_k}, \quad and \quad n = p_1^{f_1} p_2^{f_2} \cdots p_k^{f_k};$

then

$$(m, n) = p_1^{g_1} p_2^{g_2} \cdots p_k^{g_k},$$

where $g_i = min(e_i, f_i), i = 1, 2, \ldots, k.$

We will omit a formal proof, but you should have no trouble convincing yourself that it is true.

From the unique factorization theorem, it follows that each positive integer can be written in the form

$$n = p_1^{e_1} p_2^{e_2} \cdots p_k^{e_k}.$$

With a table of prime-power decompositions at hand, it is easy to find greatest common divisors. Table C (p. 218) is part of such a table: it gives the complete prime-power decompositions of some large numbers. It is included because it will be of use in some problems in other sections, and also because numbers are fascinating. If you think that eccentric opinion is peculiar to mathematicians, consider the following quotation from the book *Napoleon Bonaparte*, by J. S. C. Abbot (New York, 1904; vol. 1, chap. 10), "When he had a few moments for diversion, he not unfrequently employed them over a book of logarithms, in which he always found recreation."

Problems

1. Find the prime-power decomposition of

    (a) 111,   (b) 1234,   (c) 2345,
    (d) 3456,   (e) 4567,   (f) 111,111,
    (g) 999,999,999,999.

2. Disprove: $d \mid ab$ implies $d \mid a$ or $d \mid b$.

3. Tartaglia (1556) claimed that the sums
$$1 + 2 + 4, \qquad 1 + 2 + 4 + 8, \qquad 1 + 2 + 4 + 8 + 16, \cdots$$
are alternately prime and composite. Prove him wrong.

4. (a) DeBouvelles (1509) claimed that one or both of $6n + 1$ and $6n - 1$ are primes for all $n \geq 1$. Prove him wrong.
   (b) Show that there are infinitely many n such that both $6n - 1$ and $6n + 1$ are composite.

5. Is the difference of two consecutive cubes ever divisible by 2?

6. Prove that n is a square if and only if each exponent in its prime-power decomposition is even.

7. What is the corresponding statement for kth powers?

8. Is it possible for a prime p to divide both n and $n + 1$ $(n \geq 1)$?

9. Show that $n(n + 1)$ is never a square for $n > 0$.

10. (a) Verify that $2^5 \cdot 9^2 = 2592$.
   (b) Is $2^5 \cdot a^b = 25ab$ possible for other a, b? (Here $25ab$ denotes the digits of $2^5 \cdot a^b$ and not a product.)

11. For what primes p is $17p + 1$ a square?

12. (a) Find the smallest integer n such that $n + 1$, $n + 2$, and $n + 3$ are all composite.
   (b) If $n = 5! + 1$, show that $n + 1$, $n + 2$, $n + 3$, and $n + 4$ are composite. (For the ! notation, see Appendix B, p. 196.)
   (c) Find a sequence of 1,000 consecutive composite numbers.

13. Show that if n is composite, then $2^n - 1$ is composite.

14. Is the converse true?

15. Let p be the least prime factor of n, where n is composite. Prove that if $p > n^{1/3}$, then n/p is prime.

16. True or false? If p and q divide n, and each is greater than $n^{1/4}$, then n/pq is prime.

17. Define the *least common multiple* of a and b (written $[a, b]$) to be the smallest integer m such that $a \mid m$ and $b \mid m$.
   (a) Find $[12, 30]$ and $[pq, 2p^2]$, where p and q are distinct odd primes.
   (b) Show that $a = p_1^{e_1} p_2^{e_2} \cdots p_k^{e_k}$ and $b = p_1^{f_1} p_2^{f_2} \cdots p_k^{f_k}$ imply
   $$[a, b] = p_1^{g_1} p_2^{g_2} \cdots p_k^{g_k},$$
   where $g_i = \max(e_i, f_i)$ $(e_i \geq 0$ and $f_i \geq 0)$, $i = 1, 2, \ldots, k$.
   (c) Conclude from (b) and Theorem 3 that $[a, b] = ab/(a, b)$.

18. Fill in any missing details in the following proof of the existence of infinitely many primes. Let 2, 3, $\ldots$, p_n be all the primes. Let $N = 2 \cdot 3 \cdots p_n$, and suppose that $N = ab$. Then $a + b > p_n$, and $p_i \nmid a + b$, $i = 1, 2, \ldots, n$. Hence $a + b$ has a prime divisor greater than p_n.

19. (a) If N is odd, show that there is a square that gives a square when added to N.
   (b) If $N + a^2 = b^2$, what factors has N?

    (c) Factor 1189 by adding 1^2, 2^2, 3^3, ... until you recognize a square. (See Table B.)

    (d) Factor 9379 by the same method.

20. Establish the following test for primes. If n is odd, greater than 5, and there exist relatively prime integers a and b such that

$$a - b = n \quad \text{and} \quad a + b = p_1 p_2 \cdots p_k$$

(where p_1, p_2, ..., p_k are the odd primes less than $n^{1/2}$), then n is prime.

21. Show that the primes less than n^2 are the odd numbers not included in the arithmetic progressions

$$r^2, r^2 + 2r, r^2 + 4r, \ldots \text{(up to } n^2)$$

for $r = 3, 5, 7, \ldots$ (up to $n - 1$).

22. Let $P_n = p_1 p_2 \cdots p_n$ and $a_k = 1 + kP_n$, $k = 0, 1, \ldots, n - 1$, where the p's are the primes 2, 3, 5, 7, ... in ascending order. Show that $(a_i, a_j) = 1$ if $i \neq j$.

Linear Diophantine Equations

Consider the following variation on an old problem:

> A box contains beetles and spiders. There are 46 legs in the box; how many belong to beetles?

If we let x be the number of beetles and y the number of spiders in the box, then we know that

(1) $$6x + 8y = 46.$$

This equation has infinitely many solutions—for example,

x	-1	0	6	$4\sqrt{2}$
y	52/8	46/8	10/8	$46/8 - 3\sqrt{2}$

But none of these fit the requirements of the problem: we want x and y to be *integers*, and positive ones at that.

Equations of this sort—equations where we look for solutions in a restricted class of numbers, be they positive integers, negative integers, rational numbers, or whatever—are called *diophantine equations*. Diophantus of Alexandria (who may have lived sometime around A.D. 150) was the first to pose and solve problems that called for solutions in integers or rational numbers. Other diophantine equations we will consider in later sections include

$$x^2 + y^2 = z^2, \qquad x^2 - 2y^2 = 1, \qquad \text{and} \qquad x^4 + y^4 = z^4,$$

where we will look for solutions in integers. All of these equations have infinitely many solutions in real or complex numbers, but the third has no solutions in integers except the trivial ones where either x or y is zero. In contrast, the first and second equations both have infinitely many solutions in integers.

In this section we will consider the simplest diophantine equation: the *linear diophantine equation*

$$ax + by = c,$$

where a, b, and c are integers. For the moment, we will assume that neither a nor b is zero; we will reserve the case in which either a or b is zero until later. We want to find solutions in integers x and y. The equation $ax + by = c$ clearly has infinitely many solutions in rational numbers (and hence infinitely many solutions in real numbers), namely those given by

$$x = t, \qquad y = (c - at)/b$$

for any rational number t. But such an equation may have no solutions at all in integers. For example, $2x + 4y = 5$ has none.

Exercise 1. Why not?

With the aid of Theorems 4 and 5 of Section 1, we can find all of the integer solutions of $ax + by = c$. Before we start, let us solve the beetles and spiders problem (1) by trial. Dividing both sides of the equation by 2, we have $3x + 4y = 23$, or

$$x = \frac{23 - 4y}{3}.$$

Since x and y must be positive integers, we may let $y = 1, 2, 3, 4,$ and 5 and calculate the corresponding values of x (if $y > 5$, then x is negative):

y	1	2	3	4	5
x	19/3	5	11/3	7/3	1

Hence the diophantine equation has two solutions in positive integers: $x = 5$, $y = 2$ and $x = 1$, $y = 5$. But since the problem said that the box contained beetles (plural), we get the unique answer: 30 legs belong to spiders. Trial is sometimes the best way to solve a diophantine equation, but we want something surer.

Note that if we can find just one solution of the linear diophantine equation, then we can find infinitely many. (In keeping with our convention that lower-case italic letters denote integers unless we decree otherwise, by "solution" we mean "solution in integers.") We prove this in

LEMMA 1. *If x_0, y_0 is a solution of $ax + by = c$, then so is*

$$x_0 + bt, \qquad y_0 - at$$

for any integer t.

PROOF. We are given that $ax_0 + by_0 = c$. Thus

$$a(x_0 + bt) + b(y_0 - at) = ax_0 + abt + by_0 - bat$$
$$= ax_0 + by_0$$
$$= c,$$

so $x_0 + bt$, $y_0 - at$ satisfies the equation too. For example, we can see by inspection that

$$5x + 6y = 17$$

is satisfied by $x = 1$ and $y = 2$. It follows from Lemma 1 that $x = 1 + 6t$, $y = 2 - 5t$ is also a solution for any integer t. Thus we can write down as many solutions as we please:

t	0	1	-1	3	-5	17	-1000
x	1	7	-5	19	-29	103	-5999
y	2	-3	7	-13	27	-83	5002

Each pair x, y satisfies $5x + 6y = 17$.

Exercise 2. Find by inspection a solution of $x + 5y = 10$ and use it to write five other solutions.

After the next lemma we will see that there is no loss of generality in supposing that $(a, b) = 1$.

LEMMA 2. *If* $(a, b) \nmid c$, *then* $ax + by = c$ *has no solutions, and if* $(a, b) \mid c$, *then* $ax + by = c$ *has a solution.*

PROOF. Suppose that there are integers x_0, y_0 such that $ax_0 + by_0 = c$. Since $(a, b) \mid ax_0$ and $(a, b) \mid by_0$, it follows that $(a, b) \mid c$. Conversely, suppose that $(a, b) \mid c$. Then $c = m(a, b)$ for some m. From Theorem 4 of Section 1, we know that there are integers r and s such that

$$ar + bs = (a, b).$$

Then

$$a(rm) + b(sm) = m(a, b) = c,$$

and $x = rm$, $y = sm$ is a solution.

Exercise 3. Which of the following linear diophantine equations is impossible? (We will say that a diophantine equation is *impossible* if it has no solutions.)

(a) $14x + 34y = 90$.   (b) $14x + 35y = 91$.   (c) $14x + 36y = 93$.

Put $d = (a, b)$. Lemma 2 says that if $ax + by = c$ has a solution, then $d \mid c$. Put $a = da'$, $b = db'$, and $c = dc'$. If we divide $ax + by = c$ by d we get

$$a'x + b'y = c';$$

this equation has the same set of solutions as $ax + by = c$, and we know from Theorem 1 of Section 1 that $(a', b') = 1$. Thus, if a linear diophantine equation has solutions, then we can find them from an equation whose coefficients are relatively prime. For example, the first two equations of Exercise 3 are equivalent to

$$7x + 17y = 45 \qquad \text{and} \qquad 2x + 5y = 13,$$

and $(7, 17) = (2, 5) = 1$.

The equation $2x + 5y = 13$ has for one solution $x = 4$ and $y = 1$, and from Lemma 1 we know that $x = 4 + 5t$, $y = 1 - 2t$ is a solution for any integer t. In the next lemma, we will show that these are *all* the solutions to the equation. The problem of finding all solutions of a diophantine equation is quite distinct from the problem of finding some solutions. It is also more difficult in general. For example, the equation

$$x^3 + y^3 = z^3 + w^3$$

has solutions given by

$$x = 1 - (s - 3t)(s^2 + 3t^2),$$
$$y = -1 + (s + 3t)(s^2 + 3t^2),$$
$$z = s + 3t - (s^2 + 3t^2)^2,$$
$$w = -s + 3t + (s^2 + 3t^2)^2,$$

where s and t may be any integers. You may verify this by multiplication, if you have the patience. However, not all integer solutions are given by this formula.

LEMMA 3. *Suppose that $ab \neq 0$, $(a, b) = 1$, and x_0, y_0 is a solution of $ax + by = c$. Then all solutions of $ax + by = c$ are given by*

$$x = x_0 + bt,$$
$$y = y_0 - at,$$

where t is an integer.

PROOF. We see from Lemma 2 that the equation does have a solution, because $(a, b) = 1$ and $1 \mid c$ for all c. Then, let r, s be *any* solution of $ax + by = c$. We want to show that $r = x_0 + bt$ and $s = y_0 - at$ for some integer t. From $ax_0 + by_0 = c$ follows

$$c - c = (ax_0 + by_0) - (ar + bs)$$

or

(2) $$a(x_0 - r) + b(y_0 - s) = 0.$$

Because $a \mid a(x_0 - r)$ and $a \mid 0$, we have $a \mid b(y_0 - s)$. But we have supposed that a and b are relatively prime. It follows from Theorem 5 of Section 1 that $a \mid (y_0 - s)$. That is, there is an integer t such that

(3) $at = y_0 - s.$

Substituting in (2), this gives

$$a(x_0 - r) + bat = 0;$$

because $a \neq 0$, we may cancel it to get

(4) $x_0 - r + bt = 0.$

But (3) and (4) say that

$$s = y_0 - at,$$
$$r = x_0 + bt;$$

since r, s was *any* solution, the lemma is proved.

In Lemma 3 we assumed that $ab \neq 0$. If $ab = 0$, the problem of solving $ax + by = c$ is trivial. If $a = 0$, then x can take on any value, and y can take on one or none, according as $by = c$ has or does not have a solution in integers. The situation is similar if $b = 0$.

We can summarize the results of Lemmas 1 to 3 as follows:

THEOREM 1. *The linear diophantine equation $ax + by = c$ with $ab \neq 0$ has no solutions if $(a, b) \nmid c$. If $(a, b) \mid c$, then*

$$a'x + b'y = c',$$

where $a' = a/(a, b)$, $b' = b/(a, b)$, and $c' = c/(a, b)$, has a solution $x = r$, $y = s$, and all solutions of $ax + by == c$ are given by

$$x = r + b't,$$
$$y = s - a't,$$

where t is an arbitrary integer.

In Section 5 we will see how to solve linear diophantine equations using congruences.

The statement of the theorem may be harder to master than the process of solving linear diophantine equations. As an example, let us find all the solutions of $2x + 6y = 18$. Dividing out the common factor, we have $x + 3y = 9$. By inspection, $y = 0$, $x = 9$ is a solution. Hence all solutions are given by

(5) $x = 9 + 3t, \qquad y = -t,$

where t is an integer.

Exercise 4. Find all solutions of $2x + 6y = 20$.

Exercise 5. Find all the solutions of $2x + 6y = 18$ in *positive* integers. (Note that from (5), this is the same as asking for integers t such that $9 + 3t > 0$ and $-t > 0$.)

Problems

1. Find all integer solutions of

(a) $x + y = 2$, (c) $15x + 16y = 17$,
(b) $2x + y = 2$, (d) $15x + 18y = 17$.

2. Solve in positive integers

(a) $x + y = 2$, (c) $6x + 15y = 51$,
(b) $2x + y = 2$, (d) $7x + 15y = 51$.

3. Solve in negative integers

(a) $6x - 15y = 51$, (b) $6x + 15y = 51$.

4. Find all the positive solutions in integers of

$$x + y + z = 31,$$
$$x + 2y + 3z = 41.$$

5. Suppose that a collection of centipedes, scorpions, and worms contains 296 legs and 35 heads. How many worms are there?

6. A man bought a dozen pieces of fruit—apples and oranges—for 99 cents. If an apple costs 3 cents more than an orange and he bought more apples than oranges, how many of each did he buy?

7. A man sold his sheep for $180 each and his cows for $290 each. He received a total of $2890. How many cows did he sell?

8. How many different ways can thirty nickels, dimes, and quarters be worth $5?

9. The enrollment in a number theory class consists of sophomores, juniors, and backward seniors. If each sophomore contributes $1.25, each junior $.90, and each senior $.50, the instructor will have a fund of $25. There are 26 students; how many of each kind?

10. A says, "We three have $100 altogether." B says, "Yes, and if you had six times as much and I had one-third as much, we three would still have $100." C says, "It's not fair. I have less than $30." Who has what?

11. When Ann is half as old as Mary will be when Mary is three times as old as Mary is now, Mary will be five times as old as Ann is now. Neither Ann nor Mary may vote. How old is Ann?

12. Suppose that a, b, and c have no common factor. Show that solutions to

$$ax + by + cz = 1$$

89285

are given by

$$x = rt + crm + nb/d,$$
$$y = st + csm - na/d,$$
$$z = u - dm,$$

where m and n are arbitrary integers, r and s are such that $ar + bs = d = (a, b)$, and t and u are such that $dt + cu = 1$.

13. Apply the previous result to get solutions of

$$7x + 8y + 9z = 1.$$

14. A man cashes a check for d dollars and c cents at a bank. Assume that the teller by mistake gives the man c dollars and d cents. Assume that the man does not notice the error until he has spent 23 cents. Assume further that he then notices that he has $2d$ dollars and $2c$ cents. Assume still further that he asks you what amount the check was for. Assuming that you can accept all the assumptions, what is the answer?

15. Anna took 30 eggs to market and Barbara took 40. Each sold part of her eggs at 5 cents per egg and later sold the remainder at a lower price (in cents per egg). Each received the same amount of money. What is the smallest amount that they could have received?

Congruences

Besides being quite pretty, congruences have many applications and will be used constantly in what follows. No one who lacks an acquaintance with congruences can claim to know much about number theory. As an example of their usefulness, it is easy to show, by using congruences, that no integer of the form $8n + 7$ is a sum of three squares. We will verify this later.

We say that *a is congruent to b modulo m* (in symbols, $a \equiv b \pmod{m}$) if and only if $m \mid (a - b)$ and we will suppose always that $m > 0$.

For example, $1 \equiv 5 \pmod 4$, $-2 \equiv 9 \pmod{11}$, $6 \equiv 20 \pmod 7$, and $720 \equiv 0 \pmod{10}$.

Exercise 1. True or false? $91 \equiv 0 \pmod 7$. $3 + 5 + 7 \equiv 5 \pmod{10}$. $-2 \equiv 2 \pmod 8$. $11^2 \equiv 1 \pmod 3$.

In effect, $m \mid (a - b)$ and $a \equiv b \pmod m$ are only different notations for the same property. But a good notation can indicate new results that would be much harder to see without it. The congruence notation, which was invented by Gauss around 1800, looks a bit like the notation for equality. In fact, as we will see later, congruence and equality share many properties. Further, the notation suggests fruitful analogies with the usual algebraic operations.

There is another way to look at congruences:

THEOREM 1. $a \equiv b \pmod{m}$ *if and only if there is an integer k such that* $a = b + km$.

PROOF. Suppose that $a \equiv b \pmod m$. Then, from the definition of congruence, $m \mid (a - b)$. From the definition of divisibility, we know that since there is an integer k such that $km = a - b$, then $a = b + km$. Conversely, suppose that $a = b + km$.

Exercise 2. Complete the proof.

THEOREM 2. *Every integer is congruent (mod m) to exactly one of* 0, 1, . . . , $m - 1$.

PROOF. Write $a = qm + r$, with $0 \leq r < m$. We know from Theorem 2 of Section 1 that q and r are uniquely determined. Since $a \equiv r \pmod{m}$, the theorem is proved.

We call the number r in the last theorem the *least residue* of a (mod m). For example, the least residues of 71 modulo 2, 3, 5, 7, and 11 are 1, 2, 1, 1, and 5, respectively.

Exercise 3. To what least residue (mod 11) is each of 23, 29, 31, 37, and 41 congruent?

Yet another way of looking at congruences is given by

THEOREM 3. $a \equiv b \pmod{m}$ *if and only if a and b leave the same remainder on division by m.*

PROOF. If a and b leave the same remainder r when divided by m, then

$$a = q_1 m + r \quad \text{and} \quad b = q_2 m + r$$

for some integers q_1 and q_2. It follows that

$$a - b = (q_1 m + r) - (q_2 m + r) = m(q_1 - q_2).$$

From the definition of divisibility, we have $m \mid (a - b)$. From the definition of congruence, we conclude that $a \equiv b \pmod{m}$. To prove the converse, suppose that $a \equiv b \pmod{m}$. Then $a \equiv b \equiv r \pmod{m}$, where r is a least residue modulo m. Then from Theorem 1,

$$a = q_1 m + r \quad \text{and} \quad b = q_2 m + r$$

for some integers q_1 and q_2; since $0 \leq r < m$, the theorem is proved.

It follows from Theorems 1 and 3 that the phrases "$n \equiv 7 \pmod{8}$," "$n = 7 + 8k$ for some integer k," and "n leaves the remainder 7 when divided by 8" are different ways of saying the same thing.

Exercise 4. Say "n is odd" in three other ways.

Exercise 5. Prove that $p \mid a$ if and only if $a \equiv 0 \pmod{p}$.

Congruences share many properties with equalities, as you will show in the next three exercises.

Exercise 6. Show that

(a) $a \equiv a \pmod{m}$ for all integers a.

(b) $a \equiv b \pmod{m}$ implies $b \equiv a \pmod{m}$.

(c) $a \equiv b \pmod{m}$ and $b \equiv c \pmod{m}$ imply $a \equiv c \pmod{m}$.

Exercise 7. Show that $a \equiv b \pmod{m}$ implies $a + c \equiv b + c \pmod{m}$ for any integer c.

Exercise 8. Show that $a \equiv b \pmod{m}$ implies $ac \equiv bc \pmod{m}$ for any integer c.

Exercises 6, 7, and 8 imply that we may substitute in congruences just as we do in equations. For example, if $x \equiv 2 \pmod{5}$, then

$$2x^2 - x + 3 \equiv 2 \cdot 4 - 2 + 3 \equiv 9 \equiv 4 \pmod{5}.$$

You might try to give a detailed proof of this, noting the repeated use made of Exercises 8, 7, and 6(c). What holds in this example also holds in general.

Although $ab = ac$ and $a \neq 0$ imply $b = c$ for all integers a, b, and c, it is not true that $ab \equiv ac \pmod{m}$ and $a \not\equiv 0 \pmod{m}$ imply $b \equiv c \pmod{m}$. (The symbol $\not\equiv$ means "not congruent to.") For example,

$$3 \cdot 4 \equiv 3 \cdot 8 \pmod{12} \qquad \text{but} \qquad 4 \not\equiv 8 \pmod{12}.$$

Exercise 9. Construct a like example modulo 10.

Although we cannot cancel freely, all is not lost, as we shall see from

THEOREM 4. *If $ac \equiv bc \pmod{m}$ and $(c, m) = 1$, then $a \equiv b \pmod{m}$.*

PROOF. From the definition of congruence, $m \mid (ac - bc)$; consequently, $m \mid c(a - b)$. Because $(m, c) = 1$, we can conclude from Theorem 5 of Section 1 that $m \mid (a - b)$. That is, $a \equiv b \pmod{m}$.

Exercise 10. What values of x satisfy

(a) $2x \equiv 4 \pmod{7}$? (b) $2x \equiv 1 \pmod{7}$?

(Hint for (b): $1 \equiv 8 \pmod{7}$.)

We can, then, cancel a factor that appears on both sides of a congruence if it is relatively prime to the modulus. We now consider the case in which the factor and the modulus are not relatively prime.

THEOREM 5. *If $ac \equiv bc \pmod{m}$ and $(c, m) = d$, then $a \equiv b \pmod{m/d}$.*

PROOF. If $ac \equiv bc \pmod{m}$, then $m \mid c(a - b)$ and $m/d \mid (c/d)(a - b)$. Since we know that $(m/d, c/d) = 1$, then from Theorem 5 of Section 1 we get $m/d \mid (a - b)$, so $a \equiv b \pmod{m/d}$.

That is, we can cancel a common factor from both sides of a congruence if we divide the modulus by its greatest common divisor with the factor. For example, $30x \equiv 27 \pmod{33}$ implies $10x \equiv 9 \pmod{11}$.

Exercise 11. Which x satisfy $2x \equiv 4 \pmod 6$?

Now we can see how easy it is to show that no integer of the form $8n + 7$ is the sum of three squares. Suppose that $k = 8n + 7$ is the sum of three squares. Then $k \equiv 7 \pmod 8$ and $k = a^2 + b^2 + c^2$ for some integers a, b, and c. Thus

$$a^2 + b^2 + c^2 \equiv 7 \pmod 8.$$

We now show that this last congruence is impossible for any integers a, b, and c. What values can x^2 assume, modulo 8? Every integer has one of 0, 1, 2, 3, 4, 5, 6, and 7 for a least residue $\pmod 8$, and

$$0^2 \equiv 0, \qquad 1^2 \equiv 1, \qquad 2^2 \equiv 4, \qquad 3^2 \equiv 1, \qquad 4^2 \equiv 0, \qquad 5^2 \equiv 1,$$
$$6^2 \equiv 4, \qquad 7^2 \equiv 1,$$

all modulo 8. Thus the square of any integer is congruent modulo 8 to one of 0, 1, and 4. It is impossible to make any combination of three numbers selected from 0, 1, and 4 add up to anything congruent to 7 $\pmod 8$. (The statements $1 + 1 + 4 \equiv 6$ and $0 + 4 + 4 \equiv 8 \pmod 8$ are as close as you can come.) Hence $a^2 + b^2 + c^2$ is never congruent to 7 $\pmod 8$ for any integers a, b, and c. Thus $a^2 + b^2 + c^2 = 8n + 7$ is an impossible equation.

As another application of congruences, we will show why the process of *casting out nines* exposes errors in addition and multiplication. In case your arithmetic book never taught you to cast out nines, here is the procedure: given an addition, say

$$3141 + 5926 + 5358 = 14325,$$

add the digits in each of the addends. Thus

$$3 + 1 + 4 + 1 = 9,$$
$$5 + 9 + 2 + 6 = 22,$$
$$5 + 3 + 5 + 8 = 21.$$

If any of the sums have more than one digit, do the same thing again:

$$2 + 2 = 4, \qquad 2 + 1 = 3.$$

Eventually we will get one-digit numbers. Add them:

$$9 + 4 + 3 = 16.$$

Then apply the same procedure: $1 + 6 = 7$. We end with one digit, the magic number for the addends. In the same way, we get a magic number for the sum:

$$1 + 4 + 3 + 2 + 5 = 15 \quad \text{and} \quad 1 + 5 = 6.$$

If the addition is correct, the two magic numbers will be the same. If the magic numbers are different, as in the example, then the addition is wrong. Note that casting out nines can never ensure that a computation is correct—for example, casting out nines in

$$10 + 11 = 30$$

would reveal no error, since both magic numbers are 3. But some common errors—like failing to carry a ten—will be revealed. The rule works for multiplication too: it is easy to see that there is something wrong with

$$314 \cdot 159 = 49826$$

because the magic number of the left-hand side is 3 (the magic number for 314 is 8, and for 159 it is 6; multiplying, we see that the magic number for the product is $8 \cdot 6 = 48$ or 3). The magic number of the right-hand side is 2, since $4 + 9 + 8 + 2 + 6 = 29$. If the multiplication were correct, the two magic numbers would be the same.

Exercise 12. Cast out nines to check

(a) $123 + 456 + 789 = 1268$,
(b) $271 \cdot 828 = 224288$.

Now let us verify that the rules given above actually work. They depend on

LEMMA 1. $10^n \equiv 1 \pmod 9$ for $n = 1, 2, \ldots$.

PROOF. $10^n - 1 = 999 \cdots 99$ (n digits, all nines), which is clearly divisible by 9.

From Lemma 1 there follows

THEOREM 6. *Every integer is congruent modulo 9 to the sum of its digits.*

PROOF. Take an integer n, and let its digital representation be

$$d_k d_{k-1} d_{k-2} \cdots d_1 d_0.$$

That is,

$$n = d_k 10^k + d_{k-1} 10^{k-1} + d_{k-2} 10^{k-2} + \cdots + d_1 10^1 + d_0 10^0.$$

From Lemma 1, we see that

$$n \equiv d_k + d_{k-1} + d_{k-2} + \cdots + d_0 \pmod 9,$$

which is what we wanted to show.

For example, $3141 \equiv 3 + 1 + 4 + 1 \equiv 9 \equiv 0 \pmod 9$, and we see that 3141 is divisible by 9.

It follows that the process of casting out nines in an arithmetical operation is equivalent to considering it modulo 9. In the example

$$3141 + 5926 + 5358 = 14325,$$

Theorem 6 says that the left-hand side is congruent to

$$3 + 1 + 4 + 1 + 5 + 9 + 2 + 6 + 5 + 3 + 5 + 8 \equiv 52 \equiv 7 \pmod 9,$$

whereas the right-hand side is congruent to 6 (mod 9); if two numbers are not congruent (mod 9), then they cannot be equal. Hence 14325 is not the correct sum.

Problems

1. Prove that if $a \equiv b \pmod m$, then $a^2 \equiv b^2 \pmod m$.

2. Disprove that if $a^2 \equiv b^2 \pmod m$, then $a \equiv b \pmod m$.

3. True or false? $a \equiv b \pmod m$ implies $a^2 \equiv b^2 \pmod{m^2}$.

4. If $k \equiv 1 \pmod 4$, then what is $6k + 5$ congruent to (mod 4)?

5. Show that every prime (except 2) is congruent to 1 or 3 (mod 4).

6. Show that every prime (except 2 or 3) is congruent to 1 or 5 (mod 6).

7. What can primes (except 2, 3, or 5) be congruent to (mod 30)?

8. In the multiplication $31415 \cdot 92653 = 2910\_93995$, one digit in the product is missing and all the others are correct. What is the missing digit?

9. Show that if the last digit of n is d, then $n^2 \equiv d^2 \pmod{10}$.

10. Show that no square has as its last digit 2, 3, 7, or 8.

11. Show that no triangular number has as its last digit 2, 4, 7, or 9. (A triangular number is one of the form $n(n + 1)/2$.)

12. Prove that if $d \mid m$ and $a \equiv b \pmod m$, then $a \equiv b \pmod d$.

13. Show that the difference of two consecutive cubes is never divisible by 3.

14. Show that the difference of two consecutive cubes is never divisible by 5.

15. Prove that if the sum of the digits of an integer is divisible by 3, then the integer is divisible by 3.

16. (a) Prove that $10^k \equiv (-1)^k \pmod{11}$, $k = 0, 1, 2, \ldots$.
    (b) Conclude that
    $$d_k 10^k + d_{k-1} 10^{k-1} + \cdots + d_1 10 + d_0$$
    $$\equiv d_0 - d_1 + d_2 - d_3 + \cdots + (-1)^k d_k \pmod{11}.$$
    (c) Deduce a test for divisibility by 11.

17. A says, "27,182,818,284,590,452 is divisible by 11." B says, "No it isn't." Who is right?

18. Prove that if p is a prime and p divides no one of $a_1, a_2, \ldots, a_{p-1}$ nor any of their differences, then $a_1, a_2, \ldots, a_{p-1}$ are congruent (mod p) to 1, 2, ..., $p - 1$ in some order.

19. February 1968 had five Thursdays. What other years before 2100 will have such Februaries?

20. A *palindrome* is a number that reads the same backwards as forwards. Examples are 22, 1331, and 935686539.

(a) Prove that every four-digit palindrome is divisible by 11.
(b) What about six-digit palindromes?

21. Show that $a^5 \equiv a$ (mod 10) for all a.

22. Find an integer n such that $n \equiv 1$ (mod 2), $n \equiv 0$ (mod 3), and $n \equiv 0$ (mod 5). Can you find infinitely many?

23. Show that if $n \equiv 4$ (mod 9), then n cannot be written as the sum of three cubes.

24. Show that for $k > 0$ and $m \geq 1$, $x \equiv 1$ (mod m^k) implies $x^m \equiv 1$ (mod m^{k+1}).

25. If $n = 31,415,926,535,897$, then let

$$f(n) = 897 - 535 + 926 - 415 + 031 = 904.$$

Induce a definition for f, and prove that if $7 \mid f(n)$, then $7 \mid n$; if $11 \mid f(n)$, then $11 \mid n$; and if $13 \mid f(n)$, then $13 \mid n$. Check 118,050,660 for divisibility by 2, 3, 5, 7, 11, and 13.

Linear Congruences

After defining congruences and studying some of their properties, it is natural to look at congruences involving unknowns, like $3x \equiv 4 \pmod 5$ and $x^{17} + 3x - 3 \equiv 0 \pmod{31}$, and see how to solve them, if we can. The simplest such congruence is the *linear congruence* $ax \equiv b \pmod m$, and this is what this section is devoted to. The congruence $ax \equiv b \pmod m$ has a solution if and only if there are integers x and k such that $ax = b + km$. Hence, the problem of solving linear congruences is essentially the same as that of solving linear diophantine equations, and Theorem 1 of this section is, deep down, the same as Theorem 1 of Section 3. We are applying the same ideas in a different notation—sufficiently different that the repetition may be illuminating rather than tiresome.

We note that if there is one integer that satisfies $ax \equiv b \pmod m$, then there are infinitely many. For, suppose that $ar \equiv b \pmod m$. Then all of the integers $r + m, r + 2m, \ldots, r - m, r - 2m, \ldots$ satisfy the congruence, since

$$a(r + km) \equiv ar \equiv b \pmod m$$

for any integer k. Among the integers $r + km$, $k = 0, \pm 1, \pm 2, \ldots$, there will be exactly one—say s—that satisfies $0 \leq s < m$. This is because every integer lies between two successive multiples of m. If r satisfies the congruence and $km \leq r < (k + 1)m$ for some k, then $0 \leq r - km < m$; we can put $s = r - km$. We will single this integer out and say that by a *solution* to $ax \equiv b \pmod m$ we mean a number r such that $ar \equiv b \pmod m$ and r is a least residue $\pmod m$. For example, the congruence $2x \equiv 4 \pmod 7$ is satisfied by $x = 2, 9, 16, \ldots, -5, -12, -19, \ldots$, but all of these integers are included in the statement $x \equiv 2 \pmod 7$. They are all included in the statement $x \equiv 9 \pmod 7$ too, but we agree to call 2 *the* solution because it is a least residue modulo 7.

Unlike the familiar linear equation $ax = b$, the linear congruence $ax \equiv b$ (mod m) may have no solutions, exactly one solution, or many solutions. For example, $2x \equiv 1 \pmod 3$ is satisfied by $x = 2$ and for no other values of x that are least residues (mod 3). Hence it has just one solution, namely 2. The congruence $2x \equiv 1 \pmod 4$ has no solutions, because $4 \mid (2x - 1)$ is impossible for any x. (Since $2x - 1$ is odd, it is not a multiple of 4.) The congruence $2x \equiv 4 \pmod 6$ has two solutions, 2 and 5.

Exercise 1. Construct congruences modulo 12 with no solutions, just one solution, and more than one solution.

Exercise 2. Which congruences have no solutions?

        (a) $3x \equiv 1 \pmod{10}$,   (b) $4x \equiv 1 \pmod{10}$,

        (c) $5x \equiv 1 \pmod{10}$,   (d) $6x \equiv 1 \pmod{10}$,

        (e) $7x \equiv 1 \pmod{10}$.

Exercise 3 (optional). After Exercise 2, can you guess a criterion for telling when a congruence has no solutions?

We will now set out to prove a theorem that will enable us to examine a linear congruence and see how many solutions it has.

LEMMA 1. *If $(a, m) \nmid b$, then $ax \equiv b$ (mod m) has no solutions.*

PROOF. We will prove what is logically the same thing: if $ax \equiv b$ (mod m) has a solution, then $(a, m) \mid b$. Suppose that r is a solution. Then $ar \equiv b \pmod m$, and from the definition of congruence, $m \mid (ar - b)$, or from the definition of "divides," $ar - b = km$ for some k. Since $(a, m) \mid a$ and $(a, m) \mid km$, it follows that $(a, m) \mid b$.

For example, $6x \equiv 7 \pmod 8$ has no solutions.

LEMMA 2. *If $(a, m) = 1$, then $ax \equiv b$ (mod m) has exactly one solution.*

PROOF. Since $(a, m) = 1$, we know that there are integers r and s such that $ar + ms = 1$. Multiplying by b gives

$$a(rb) + m(sb) = b.$$

We see that $arb - b$ is a multiple of m, or

$$a(rb) \equiv b \pmod m.$$

The least residue of rb modulo m is then a solution of the linear congruence.

It remains to show that there is not more than one solution. Suppose that both r and s are solutions. That is, since

$$ar \equiv b \pmod{m} \qquad \text{and} \qquad as \equiv b \pmod{m},$$

then $ar \equiv as \pmod{m}$. Because $(a, m) = 1$, we can apply Theorem 4 of the last section, cancel the common factor and get $r \equiv s \pmod{m}$. That is, $m \mid (r - s)$. But r and s are least residues $\pmod{m}$, so

$$0 \leq r < m \qquad \text{and} \qquad 0 \leq s < m.$$

Thus $-m < r - s < m$; together with $m \mid (r - s)$, this gives $r - s = 0$, or $r = s$, and the solution is unique. The above argument is quite general and will be used often: if two least residues $\pmod{m}$ are congruent $\pmod{m}$, then they are equal.

Inspection is one way of solving congruences with small moduli, and another is substituting all possible values for the variable. But the best way is to manipulate the coefficients until cancellation is possible. For example, to solve $4x \equiv 1 \pmod{15}$, we can write

$$4x \equiv 1 \equiv 16 \pmod{15}$$

and cancel a 4 to get $x \equiv 4 \pmod{15}$. As another example, let us solve $14x \equiv 27 \pmod{31}$. From

$$14x \equiv 27 \equiv 58 \pmod{31}$$

we get $7x \equiv 29 \pmod{31}$. We continue adding 31 until we can cancel the 7:

$$7x \equiv 29 \equiv 60 \equiv 91 \pmod{31},$$

so we get $x \equiv 13 \pmod{31}$, and 13 is the solution.

This method is the best to use when solving linear diophantine equations. The equation $ax + by = c$ implies the two congruences

$$ax \equiv c \pmod{b} \qquad \text{and} \qquad by \equiv c \pmod{a}.$$

We can choose either one, solve for the variable, and then substitute the result into the original equation to get all the solutions. For example, let us solve $9x + 16y = 35$. This gives $16y \equiv 35 \pmod{9}$ or, manipulating the coefficients, $7y \equiv 35 \pmod{9}$, from which we get $y \equiv 5 \pmod{9}$. That is, $y = 5 + 9t$ for some integer t. Substituting this in the original equation, we get

$$9x + 16(5 + 9t) = 35,$$

or $9x + 144t = -45$, or $x + 16t = -5$. We thus have all the solutions:

$$x = -5 - 16t,$$
$$y = 5 + 9t,$$

t an integer.

Exercise 4. Solve

$$\text{(a) } 8x \equiv 1 \text{ (mod 15)}, \qquad \text{(b) } 9x + 10y = 11.$$

We now consider the case where a and m are not necessarily relatively prime in

LEMMA 3. *If $(a, m) \mid b$, then $ax \equiv b \, (mod \, m)$ has exactly (a, m) solutions.*

PROOF. We will construct (a, m) solutions of the congruence and then show that there are no others. Let $d = (a, m)$, and put

$$a = da', \qquad b = db', \qquad m = dm'.$$

From Theorem 5 of Section 4, we have

(1) $\qquad a'x \equiv b' \text{ (mod } m') \qquad$ and $\qquad (a', m') = 1,$

the latter statement following from Theorem 1 of Section 1. Lemma 2 applies to the congruence in (1) and tells us that it has just one solution. Call the solution r. We claim that the d numbers

(2) $\qquad r, r + m', r + 2m', \ldots, r + (d - 1)m'$

are all the solutions of $ax \equiv b$ (mod m). First, each of these satisfies the congruence, because for $k = 1, 2, \ldots, d - 1$, we have

$$a(r + km') = a'dr + a'dkm' = a'rd + a'k(m'd).$$

Since $a'r \equiv b'$ (mod m') and $m'd = m$, we have

$$a'rd + a'k(m'd) \equiv b'd + a'km \equiv b'd \equiv b \text{ (mod } m).$$

That is,

$$a(r + km') \equiv b \text{ (mod } m).$$

Second, each of the numbers in (2) is a least residue (mod m), because for $k = 0, 1, \ldots, d - 1$,

$$0 \le r + km' \le r + (d - 1)m' < m' + (d - 1)m' = dm' = m.$$

Third, no two of the numbers in (2) are congruent (mod m), because they are distinct least residues (mod m). Thus we have shown that $ax \equiv b$ (mod m) has (a, m) solutions.

It remains to show that there are no others. Let r be the solution in (2), and let s be any solution of $ax \equiv b$ (mod m). We want to show that s is one of the numbers in (2). We have

$$ar \equiv as \equiv b \text{ (mod } m).$$

It follows from Theorem 5 of Section 4 that $r \equiv s \pmod{m'}$. That is, $s - r = km'$, or

$$s = r + km',$$

for some k. But s is a least residue $\pmod{m}$, and all of the least residues $\pmod{m}$ of the form $r + km'$ for some k appear in (2). Hence s is one of the numbers in (2). Since s was any solution, it follows that the solutions in (2) are all of the solutions, and the lemma is proved.

Let us look at an example. Consider $6x \equiv 15 \pmod{33}$; Lemma 3 says that the congruence has exactly three solutions. Cancelling a 3, we get $2x \equiv 5 \pmod{11}$, and solving by the usual procedure,

$$2x \equiv 5 \equiv 16 \pmod{11} \quad \text{and} \quad x \equiv 8 \pmod{11}.$$

That is, $6x \equiv 15 \pmod{33}$ is satisfied by any $x \equiv 8 \pmod{11}$. The integers included in the last congruence which are least residues $\pmod{33}$ are 8, 19, and 30, and these are the three solutions.

Exercise 5. Determine the number of solutions of each of the following congruences:

$$3x \equiv 6 \pmod{15}, \qquad 4x \equiv 8 \pmod{15}, \qquad 5x \equiv 10 \pmod{15},$$
$$6x \equiv 11 \pmod{15}, \qquad 7x \equiv 14 \pmod{15}.$$

Exercise 6. Find all of the solutions of $5x \equiv 10 \pmod{15}$.

We can summarize the results of Lemmas 1 to 3 in

THEOREM 1. $ax \equiv b \pmod{m}$ *has no solutions if* $(a, m) \nmid b$. *If* $(a, m) \mid b$, *then there are exactly* (a, m) *solutions.*

Exercise 7. Solve the rest of the congruences in Exercise 5.

This completes the analysis of a single linear congruence. In the remainder of the section, we will consider a special kind of system of linear congruences, and prove the Chinese Remainder Theorem, which is important theoretically. The name of the theorem comes from the inclusion in some old Chinese manuscripts of problems like "Find a number that leaves a remainder of 2 when divided by 3, of 4 when divided by 5, and of 6 when divided by 7." In our notation, the problem is to find x such that

$$x \equiv 2 \pmod{3}, \qquad x \equiv 4 \pmod{5}, \qquad \text{and} \qquad x \equiv 6 \pmod{7}.$$

Exercise 8. Verify that 104 satisfies the problem posed above.

Exercise 9 (optional). Find infinitely many other solutions.

THEOREM 2 (the Chinese Remainder Theorem). *The system of congruences*

(4) $x \equiv a_i \ (mod \ m_i), \qquad i = 1, 2, \ldots, k,$

where $(m_i, m_j) = 1$ if $i \neq j$, has a unique solution modulo $m_1 m_2 \cdots m_k$.

Before we prove the theorem, we will consider an example. It includes the idea of the proof and shows how to go about finding the unique solution in practice. Let us look for x satisfying

$x \equiv 1 \ (\text{mod } 3), \qquad x \equiv 2 \ (\text{mod } 5), \qquad \text{and} \qquad x \equiv 3 \ (\text{mod } 7).$

The first congruence gives $x = 1 + 3k_1$ for some k_1. Substituting this into the second congruence, we see that k_1 must satisfy

$$1 + 3k_1 \equiv 2 \ (\text{mod } 5);$$

consequently, $k_1 \equiv 2 \ (\text{mod } 5)$. That is, $k_1 = 2 + 5k_2$ for some k_2, and thus

$$x = 1 + 3k_1 = 1 + 3(2 + 5k_2) = 7 + 15k_2$$

satisfies the first two congruences. If, in addition, x satisfies the third, we must have

$$7 + 15k_2 \equiv 3 \ (\text{mod } 7),$$

which implies $k_2 \equiv 3 \ (\text{mod } 7)$. Thus

$$x = 7 + 15(3 + 7k_3) = 52 + 105k_3$$

satisfies all three congruences for any integer k_3. Otherwise expressed, any $x \equiv 52 \ (\text{mod } 105)$ satisfies the three congruences. In fact, 52 is the unique solution modulo 105.

PROOF OF THEOREM 2. We first show, by induction, that system (4) has a solution. The result is obvious when $k = 1$. Let us consider the case $k = 2$. If $x \equiv a_1 \ (\text{mod } m_1)$, then $x = a_1 + k_1 m_1$ for some k_1. If in addition $x \equiv a_2 \ (\text{mod } m_2)$, then

$$a_1 + k_1 m_1 \equiv a_2 \ (\text{mod } m_2)$$

or

$$k_1 m_1 \equiv a_2 - a_1 \ (\text{mod } m_2).$$

Because $(m_2, m_1) = 1$, we know that this congruence, with k_1 as the unknown, has a unique solution modulo m_2. Call it t. Then $k_1 = t + k_2 m_2$ for some k_2, and

$$x = a_1 + (t + k_2 m_2)m_1 \equiv a_1 + t m_1 \ (\text{mod } m_1 m_2)$$

satisfies both congruences.

Suppose that system (4) has a solution (mod $m_1 m_2 \cdots m_k$) for $k = r - 1$. Then there is a solution, s, to the system

$$x \equiv a_i \pmod{m_i}, \qquad i = 1, 2, \ldots, r - 1.$$

But the system

$$x \equiv s \pmod{m_1 m_2 \cdots m_{r-1}},$$
$$x \equiv a_r \pmod{m_r}$$

has a solution modulo the product of the moduli, just as in the case $k = 2$, because $(m_1 m_2 \cdots m_{k-1}, m_k) = 1$ (this statement is true because no prime that divides m_i, $i = 1, 2, \ldots, k - 1$, can divide m_k).

It is easy to see that the solution is unique. If r and s are both solutions of the system, then

$$r \equiv s \equiv a_i \pmod{m_i}, \qquad i = 1, 2, \ldots, k,$$

so $m_i \mid (r - s)$, $i = 1, 2, \ldots, k$. Thus $r - s$ is a common multiple of $m_1, m_2, \ldots, m_k$, and because the moduli are relatively prime in pairs, we have $m_1 m_2 \cdots m_k \mid (r - s)$. But since r and s are least residues modulo $m_1 m_2 \cdots m_k$,

$$-m_1 m_2 \cdots m_k < r - s < m_1 m_2 \cdots m_k,$$

whence $r - s = 0$.

Problems

1. Solve the following congruences.

(a) $2x \equiv 1 \pmod{17}$,
(b) $3x \equiv 1 \pmod{17}$,
(c) $3x \equiv 6 \pmod{18}$,

(d) $4x \equiv 6 \pmod{18}$,
(e) $40x \equiv 191 \pmod{6191}$.

2. Construct linear congruences modulo 20 with no solutions, just one solution, and more than one solution. Can you find one with 20 solutions?

3. What possibilities are there for the number of solutions of a linear congruence (mod 20)?

4. Solve the systems of congruences

(a) $x \equiv 1 \pmod 2$, $x \equiv 1 \pmod 3$;
(b) $x \equiv 1 \pmod 2$, $x \equiv 1 \pmod 3$, $x \equiv 6 \pmod 7$;
(c) $x \equiv 3 \pmod 5$, $x \equiv 5 \pmod 7$, $x \equiv 7 \pmod{11}$;
(d) $2x \equiv 1 \pmod 5$, $3x \equiv 2 \pmod 7$, $4x \equiv 1 \pmod{11}$;
(e) $x \equiv 31 \pmod{41}$, $x \equiv 59 \pmod{26}$.

5. Solve $9x \equiv 4 \pmod{2401}$.

6. When the marchers in the annual Mathematics Department Parade lined up 4 abreast, there was one odd man; when they tried 5 in a line, there were two left over; and when 7 abreast, there were three left over. How large is the department?

7. Find the smallest odd n, $n > 3$, such that $3 \mid n$, $5 \mid n + 2$, and $7 \mid n + 4$.

8. Find a number such that half of it is a square, a third of it is a cube, and a fifth of it is a fifth power.

9. Find a multiple of 7 that leaves the remainder 1 when divided by 2, 3, 4, 5, or 6.

10. The three consecutive integers 48, 49, and 50 each have a square factor.

   (a) Find n such that $3^2 \mid n$, $4^2 \mid n + 1$, and $5^2 \mid n + 2$.
   (b) Can you find n such that $2^2 \mid n$, $3^2 \mid n + 1$, and $4^2 \mid n + 2$?

11. If $x \equiv r \pmod{m}$ and $x \equiv s \pmod{m + 1}$, show that

$$x \equiv r(m + 1) - sm \pmod{m(m + 1)}.$$

12. Solve for x and y.

   (a) $x + 2y \equiv 3 \pmod{7}$, $3x + y \equiv 2 \pmod{7}$.
   (b) $x + 2y \equiv 3 \pmod{6}$, $3x + y \equiv 2 \pmod{6}$.

13. What three positive integers upon being multiplied by 3, 5, and 7 respectively and the products divided by 20 have remainders in arithmetic progression with common difference 1 and quotients equal to remainders?

14. Consider the system

$$x \equiv a_i \pmod{m_i}, \qquad i = 1, 2, \ldots, k,$$

where the moduli are relatively prime in pairs. Let

$$M_i = (m_1 m_2 \cdots m_k)/m_i, \qquad i = 1, 2, \ldots, k.$$

Let s_i denote the solution of $M_i x \equiv 1 \pmod{m_i}$, $i = 1, 2, \ldots, k$. Show that

$$s = a_1 s_1 M_1 + a_2 s_2 M_2 + \cdots + a_k s_k M_k$$

satisfies each of the congruences in the system.

Carry out the above process to solve the system in Problem 4(b).

15. Find the smallest integer n, $n > 2$, such that

$$2 \mid n, \quad 3 \mid n + 1, \quad 4 \mid n + 2, \quad 5 \mid n + 3, \quad \text{and} \quad 6 \mid n + 4.$$

16. Suppose that the moduli in the system

$$x \equiv a_i \pmod{m_i}, \qquad i = 1, 2, \ldots, k$$

are not relatively prime in pairs. Find a condition that the a_i must satisfy in order that the system have a solution.

17. How many multiples of b are there in the sequence

$$a, 2a, 3a, \ldots, ba?$$

Fermat's and Wilson's Theorems

In this section we will prove

THEOREM 1 (Fermat's Theorem). *If p is prime and* $(a, p) = 1$, *then*

$$a^{p-1} \equiv 1 \ (mod \ p).$$

This theorem, which was stated without proof by Fermat in 1640, was fundamental to the progress of number theory. It is vital, as we shall see, to the study of quadratic congruences, and it has many other applications. Its statement and proof are simple, but its effects are great. We will also prove

THEOREM 2 (Wilson's Theorem). *p is a prime if and only if*

$$(p - 1)! \equiv -1 \ (mod \ p).$$

(*For the* ! *notation see Appendix B, p.* 196.)

The method used for proving this theorem is close to that of Fermat's Theorem, and it is also helpful in the study of quadratic congruences. (Wilson's Theorem is not really Wilson's. He guessed that it was true and wrote about it to the mathematician Waring, who published it without proof in 1770. Wilson was not the first to guess it, though—Leibnitz had also discovered it in 1682—and the first proof was given by Lagrange, very shortly after Waring's announcement.) Wilson's Theorem is remarkable because it gives a condition both necessary and sufficient for a number to be prime. Thus, in theory, the problem of determining whether a given number is prime is completely solved. But for large integers, the computational difficulties are great. For the moderate-sized prime

$$p = 162,259,276,829,213,363,391,578,010,288,127,$$

the calculation of the least residue of $(p - 1)!$ (mod p) would take about 10^{33} multiplications of two 33-digit numbers, followed by division by p. Even our fastest computers are not fast enough. Compare, for example, the calculation of 12! (mod 13) with the labor in verifying that 13 is divisible by neither 2 nor 3.

We start the proof of Fermat's Theorem with

LEMMA 1. *If $(a, m) = 1$, then the least residues of*

(1) $$a, 2a, 3a, \ldots, (m - 1)a \ (mod \ m),$$

are

(2) $$1, 2, 3, \ldots, m - 1$$

in some order.

Stated differently, if $(a, m) = 1$, then each integer is congruent (mod m) to exactly one of $0, a, 2a, \ldots, (m - 1)a$. For example, take $m = 8$ and $a = 3$: the numbers in (1) are then

$$3, 6, 9, 12, 15, 18, 21,$$

and their least residues (mod 8) are

$$3, 6, 1, 4, 7, 2, 5.$$

PROOF OF LEMMA 1. There are $m - 1$ numbers in (1), none congruent to 0 (mod m). Hence each of them is congruent (mod m) to one of the numbers in (2). If we show that no two of the integers in (1) are congruent (mod m), then it follows that their least residues (mod m) are all different, and hence are a permutation of $1, 2, \ldots, m - 1$. Suppose that two of the integers in (1) are congruent (mod m): that is,

$$ra \equiv sa \ (mod \ m);$$

because $(a, m) = 1$ we can cancel (Theorem 4 of Section 4) and get

$$r \equiv s \ (mod \ m).$$

But r and s are least residues modulo m; by an argument we have used several times before, it follows that $r = s$. This proves the lemma.

PROOF OF FERMAT'S THEOREM. Given any prime p, Lemma 1 says that if $(a, p) = 1$, then the least residues of

$$a, 2a, \ldots, (p - 1)a \ (mod \ p)$$

are a permutation of

$$1, 2, \ldots, p - 1.$$

Hence their products are congruent (mod p):

$$a \cdot 2a \cdot 3a \cdots (p-1)a \equiv 1 \cdot 2 \cdot 3 \cdots (p-1) \ (\text{mod } p),$$

or

$$a^{p-1}(p-1)! \equiv (p-1)! \ (\text{mod } p).$$

Since p and $(p-1)!$ are relatively prime, the last congruence gives

$$a^{p-1} \equiv 1 \ (\text{mod } p),$$

which is Fermat's Theorem.

Exercise 1. Verify that the theorem is true for $a = 2$ and $p = 5$.

Fermat's Theorem is sometimes stated in a slightly different way:

COROLLARY. If p is a prime, then

$$a^p \equiv a \ (\text{mod } p)$$

for all a.

PROOF. If $(a, p) = 1$, this follows from Fermat's Theorem. If $(a, p) = p$, then the corollary says $0 \equiv 0 \ (\text{mod } p)$, which is true. There are no other cases.

As an example, let us verify that $3^{16} \equiv 1 \ (\text{mod } 17)$. It is not necessary to calculate the large integer 3^{16} and then divide it by 17: we can proceed in stages, reducing modulo 17 as we go. We have

$$3^3 \equiv 27 \equiv 10 \ (\text{mod } 17).$$

Squaring, we get

$$3^6 \equiv 100 \equiv -2 \ (\text{mod } 17);$$

squaring again yields $3^{12} \equiv 4 \ (\text{mod } 17)$. Thus

$$3^{16} \equiv 3^{12} \cdot 3^3 \cdot 3 \equiv 4 \cdot 10 \cdot 3 \equiv 120 \equiv 1 \ (\text{mod } 17).$$

Exercise 2. Calculate 2^5 and 2^{10} (mod 11).

To prove Wilson's Theorem, we need two lemmas:

LEMMA 2. $x^2 \equiv 1 \ (\text{mod } p)$ *has exactly two solutions:* 1 *and* $p - 1$.

PROOF. Let r be any solution of $x^2 \equiv 1 \ (\text{mod } p)$. By *solution* we mean, as we did for linear congruences, a least residue that satisfies the congruence. Then $r^2 - 1 \equiv 0 \ (\text{mod } p)$, so

$$p | (r+1)(r-1).$$

Hence $p|(r + 1)$ or $p|(r - 1)$; otherwise expressed,

$$r + 1 \equiv 0 \quad \text{or} \quad r - 1 \equiv 0 \,(\text{mod}\,p),$$

so $r \equiv p - 1$ or $1 \,(\text{mod}\,p)$. Since r is a least residue $(\text{mod}\,p)$, it follows that $r = p - 1$ or 1. It is easy to verify that both of these numbers actually satisfy $x^2 \equiv 1 \,(\text{mod}\,p)$.

Lemma 2 has a familiar analogy: $x^2 = 1$ is satisfied only when $x = 1$ or $x = -1$, and $x^2 \equiv 1 \,(\text{mod}\,p)$ is satisfied only when $x \equiv 1 \,(\text{mod}\,p)$ or $x \equiv -1$ $(\text{mod}\,p)$.

LEMMA 3. *Let p be an odd prime, and let a' denote the solution of $ax \equiv 1$ (mod p), $a = 1, 2, \ldots, p - 1$. (That is,*

$$aa' \equiv 1 \,(mod\,p) \quad and \quad 0 \le a' < p.)$$

Then

$$a' \not\equiv b' \,(mod\,p) \quad if \quad a \not\equiv b \,(mod\,p), \quad and$$

$$a' \equiv a \,(mod\,p) \quad only\ when \quad a = 1 \quad or \quad a = p - 1.$$

PROOF. First we note that since $(a, p) = 1$, a' exists and is uniquely determined, because $ax \equiv 1 \,(\text{mod}\,p)$ has exactly one solution. Suppose that $a' \equiv b' \,(\text{mod}\,p)$. Then

$$1 \equiv aa' \equiv ab' \,(\text{mod}\,p),$$

which implies

$$b \equiv ab'b \equiv a \,(\text{mod}\,p),$$

and this proves the first assertion of the lemma. For the second assertion, suppose that $a \equiv a' \,(\text{mod}\,p)$. Then

$$1 \equiv aa' \equiv a^2 \,(\text{mod}\,p),$$

and from Lemma 2 we know that this is possible only if $a = 1$ or $a = p - 1$.

As an illustration of this result, let us take $p = 13$. We have

a	1	2	3	4	5	6	7	8	9	10	11	12
a'	1	7	9	10	8	11	2	5	3	4	6	12
aa'	1	14	27	40	40	66	14	40	27	40	66	144

The set of numbers in the second line is a permutation of the set of numbers in the first, $aa' \equiv 1 \,(\text{mod}\,13)$ in every case, and $a \equiv a' \,(\text{mod}\,13)$ only when $a = 1$ or 12.

PROOF OF WILSON'S THEOREM. Note that $(2 - 1)! \equiv -1 \pmod 2$. Thus the theorem is true if $p = 2$, and for the rest of the proof we can assume that p is an odd prime. From Lemma 3, we know that we can separate the numbers

$$2, 3, \ldots, p - 2$$

into $(p - 3)/2$ pairs such that each pair consists of an integer a and its associated a', which is different from a. For example, for $p = 13$ the pairs are

$$(2, 7), (3, 9), (4, 10), (5, 8), (6, 11).$$

Exercise 3. What are the pairs when $p = 11$?

The product of the two integers in each pair is congruent to 1 (mod p), so it follows that

$$2 \cdot 3 \cdots (p - 2) \equiv 1 \pmod p.$$

Hence

$$(p - 1)! \equiv 1 \cdot 2 \cdot 3 \cdots (p - 2)(p - 1) \equiv 1 \cdot 1 \cdot (p - 1) \equiv -1 \pmod p,$$

and we have proved half of the theorem. It remains to prove the other half and show that if

$$(3) \qquad\qquad (n - 1)! \equiv -1 \pmod n,$$

then n is a prime. Suppose that $n = ab$ for some integers a and b, with $a \neq n$. From (3), we have

$$n \mid (n - 1)! + 1,$$

and since $a \mid n$, we have

$$(4) \qquad\qquad a \mid (n - 1)! + 1.$$

But since $a \leq n - 1$, it follows that one of the factors of $(n - 1)!$ is a itself. Thus

$$(5) \qquad\qquad a \mid (n - 1)!.$$

But (4) and (5) imply that $a \mid 1$. Hence the only positive divisors of n are 1 and n, and thus n is a prime.

Exercise 4. Verify that $(p - 1)! \equiv -1 \pmod p$ for $p = 3, 5$, and 7.

Problems

1. What is the remainder when 314^{159} is divided by 7?

2. What is the remainder when 314^{162} is divided by 163?

3. What is the last digit of 7^{355}?

4. What are the last two digits of 7^{355}?

5. Show that
$$(p - 1)(p - 2)\cdots(p - r) \equiv (-1)^r r! \pmod{p},$$
for $r = 1, 2, \ldots, p - 1$.

6. Note that
$$6! \equiv -1 \pmod{7},$$
$$5!1! \equiv 1 \pmod{7},$$
$$4!2! \equiv -1 \pmod{7},$$
$$3!3! \equiv 1 \pmod{7}.$$

Try the same sort of calculation (mod 11).

7. Guess a theorem from the data of Problem 6, and prove it.

8. (a) Calculate $(n - 1)! \pmod{n}$ for $n = 4, 6, 8, 9$, and 10.
(b) Guess and prove a theorem.

9. Show that
$$2(p - 3)! + 1 \equiv 0 \pmod{p}$$
if p is an odd prime greater than 5.

10. (a) Prove that if $r! \equiv (-1)^r \pmod{p}$, then
$$(p - r - 1)! \equiv -1 \pmod{p}.$$
(b) Find an example of such a p and r.

11. (a) Show that
$$(k + 1)^p - k^p \equiv 1 \pmod{p},$$
for $k = 0, 1, \ldots$.
(b) Derive Fermat's Theorem from this.

12. In 1732 Euler said, "I derived [certain] results from the elegant theorem, of whose truth I am certain, although I have no proof: $a^n - b^n$ is divisible by the prime $n + 1$ if neither a nor b is." Prove this theorem, using Fermat's Theorem.

13. Suppose that p is an odd prime.
(a) Show that
$$1^{p-1} + 2^{p-1} + \cdots + (p - 1)^{p-1} \equiv -1 \pmod{p}.$$
(b) Show that
$$1^p + 2^p + \cdots + (p - 1)^p \equiv 0 \pmod{p}.$$

(c) Prove that if $2^m \not\equiv 1 \pmod{p}$, then

$$1^m + 2^m + \cdots + (p-1)^m \equiv 0 \pmod{p}.$$

14. Leo Moser has proved—in *Scripta Mathematica*, vol. 22 (1956), p. 288—a theorem from which Fermat's Theorem and part of Wilson's Theorem can be deduced: if p is prime, then

$$a^p(p-1)! \equiv a(p-1) \pmod{p}$$

for all a. Show that this implies Fermat's Theorem and that

$$(p-1)! \equiv -1 \pmod{p}.$$

15. Show that the converse of Fermat's Theorem is false by calculating 2^{340} (mod 341) and noting that $341 = 11 \cdot 31$.

16. A composite n such that $n \mid (2^n - 2)$ is called a *pseudoprime*. There are infinitely many, and the smallest two are 341 and 561. Verify that 561 is a pseudoprime.

17. A composite n such that $n \mid (a^n - a)$ for all a is called an *absolute pseudoprime*. The smallest absolute pseudoprime is 561. Show that 341 is not an absolute pseudoprime by verifying that $341 \nmid (11^{341} - 11)$.

18. Show that for any two different primes p, q,

(a) $pq \mid (a^{p+q} - a^{p+1} - a^{q+1} + a^2)$ for all a.
(b) $pq \mid (a^{pq} - a^p - a^q + a)$ for all a.

19. Show that if p is an odd prime, then $2p \mid (2^{2p-1} - 2)$.

20. If p is an odd prime and $(a, p) = 1$, then what values can $a^{(p-1)/2}$ assume, modulo p?

21. For what n is it true that

$$p \mid (1 + n + n^2 + \cdots + n^{p-2})?$$

22. Show that every odd prime except 5 divides some number of the form $111 \ldots 11$ (k digits, all ones).

23. If p is an odd prime, and if $(a, p) = 1$, $n \mid p - 1$, and $a \equiv c^n \pmod{p}$, prove that

$$p \mid (a^{(p-1)/n} - 1).$$

The Divisors of an Integer

It would be natural now to continue studying congruences by taking up quadratic congruences, but partly for the sake of variety, we will take up a different subject and return to congruences later.

Let n be a positive integer. Let $d(n)$ denote the number of positive divisors of n (including 1 and n), and let $\sigma(n)$ denote their sum. That is,

$$d(n) = \sum_{d|n} 1 \quad \text{and} \quad \sigma(n) = \sum_{d|n} d.$$

(This notation may be unfamiliar. If so, now is the time to look at Appendix B, p. 196.) These functions occur frequently, and in this section we will derive some of their properties. In the next section we will use them to study perfect numbers, a subject that engaged the ancient Greek mathematicians and has received constant attention ever since.

Exercise 1. Verify that the following table is correct as far as it goes, and complete it.

n	1	2	3	4	5	6	7	8	9	10	11	12	13	14	15	16
$d(n)$	1	2	2	3	2	4	2	4	3	4						

If p is a prime, then $d(p) = 2$, because the only positive divisors of p are 1 and p. Since p^2 has divisors 1, p, and p^2, then $d(p^2) = 3$.

Exercise 2. What is $d(p^3)$? Generalize to $d(p^n)$, $n = 4, 5, \ldots$.

If $p \neq q$, then pq has divisors 1, p, q, and pq, so $d(pq) = 4$. (In this section, as elsewhere, p and q will stand for primes.) Similarly, the divisors of p^2q are 1, p, p^2, q, pq, and p^2q, so $d(p^2q) = 6$.

Exercise 3. What is $d(p^3q)$? What is $d(p^nq)$ for any positive n?

After Exercises 2 and 3, you may have guessed

THEOREM 1. *If $p_1^{e_1}p_2^{e_2} \cdots p_k^{e_k}$ is the prime-power decomposition of n (recall that this means that $e_i \geq 1$ for all i and $p_i \neq p_j$ if $i \neq j$), then*

$$d(n) = (e_1 + 1)(e_2 + 1)(e_3 + 1) \cdots (e_k + 1).$$

Expressed in the more compact product notation, we could write

$$\text{if} \quad n = \prod_{i=1}^{k} p_i^{e_i}, \quad \text{then} \quad d(n) = \prod_{i=1}^{k} (e_i + 1)$$

or even

$$\text{if} \quad n = \prod_{p|n} p^{a_p}, \quad \text{then} \quad d(n) = \prod_{p|n} (a_p + 1).$$

PROOF. Let D denote the set of numbers

(1) $$p_1^{f_1}p_2^{f_2} \cdots p_k^{f_k}, \quad 0 \leq f_i \leq e_i.$$

We claim that D is exactly the set of divisors of n. First, we note that every number in the set is a divisor of n, because we can find for each number in (1) an integer whose product with the number is n; namely,

$$p_1^{e_1-f_1}p_2^{e_2-f_2} \cdots p_k^{e_k-f_k}.$$

Second, suppose that d is a divisor of n. If $p|d$, then $p|n$, so each prime in the prime-power decomposition of d must appear in the prime-power decomposition of n. Thus

$$d = p_1^{f_1}p_2^{f_2} \cdots p_k^{f_k},$$

where some (or all) of the exponents may be zero. Moreover, no exponent f_i is larger than e_i. (If it were, we would have a situation in which $p_i^{f_i}|d$ and $d|n$, which implies $p_i^{f_i}|n$. This is impossible if $f_i > e_i$.) Thus every divisor of n is a member of the set D. Thus D is identical with the set of divisors of n.

Each f_i in (1) may take on $e_i + 1$ values. Thus there are

$$(e_1 + 1)(e_2 + 1)(e_3 + 1) \cdots (e_k + 1)$$

numbers in D, and because of the unique factorization theorem, they are all different. This proves the theorem.

Exercise 4. Calculate $d(240)$.

Now we will get a formula for $\sigma(n)$.

Exercise 5. Verify that the following table is correct as far as it goes, and complete it.

n	1	2	3	4	5	6	7	8	9	10	11	12	13	14
$\sigma(n)$	1	3	4	7	6	12	8	15						

As with $d(n)$, some special cases are easy. For example, $\sigma(p) = p + 1$ for all primes p. Furthermore, the divisors of p^2 are 1, p, and p^2, so $\sigma(p^2) = 1 + p + p^2 = (p^3 - 1)/(p - 1)$.

Exercise 6. What is $\sigma(p^3)$? $\sigma(pq)$, where p and q are different primes?

Exercise 7. Show that $\sigma(2^n) = 2^{n+1} - 1$.

Exercise 8. What is $\sigma(p^n)$, $n = 1, 2, \ldots$?

Let us calculate $\sigma(p^e q^f)$, where p and q are different primes, and see if it suggests a general result. The divisors of $p^e q^f$ are

$$
\begin{array}{ccccc}
1 & p & p^2 & \cdots & p^e, \\
q & pq & p^2 q & \cdots & p^e q, \\
q^2 & pq^2 & p^2 q^2 & \cdots & p^e q^2, \\
& & \cdots & & \\
q^f & pq^f & p^2 q^f & \cdots & p^e q^f.
\end{array}
$$

If we add across each row, we get

$$
\begin{aligned}
\sigma(p^e q^f) &= (1 + p + \cdots + p^e) + q(1 + p + \cdots + p^e) \\
&\quad + q^2(1 + p + \cdots + p^e) + \cdots \\
&\quad + q^f(1 + p + \cdots + p^e) \\
&= (1 + q + \cdots + q^f)(1 + p + \cdots + p^e).
\end{aligned}
$$

Or, summing the geometric series,

$$
\sigma(p^e q^f) = \frac{p^{e+1} - 1}{p - 1} \cdot \frac{q^{f+1} - 1}{q - 1}.
$$

Together with

$$
\sigma(p^e) = \frac{p^{e+1} - 1}{p - 1},
$$

this might suggest

THEOREM 2. *If the prime-power decomposition of n is $p_1^{e_1} p_2^{e_2} \cdots p_k^{e_k}$, then*

(2) $$\sigma(n) = \frac{p_1^{e_1+1} - 1}{p_1 - 1} \cdot \frac{p_2^{e_2+1} - 1}{p_2 - 1} \cdots \frac{p_k^{e_k+1} - 1}{p_k - 1}.$$

PROOF. We want to show that the sum of the divisors of n is

(3) $(1 + p_1 + \cdots + p_1^{e_1})(1 + p_2 + \cdots + p_2^{e_2}) \cdots (1 + p_k + \cdots + p_k^{e_k}),$

for (3) is the same as the right-hand side of (2). Each term in the sum (3), after the parentheses have been removed by repeated multiplication, is a product of k factors: one from the first parenthesis, one from the second, and so on. Thus each term has the form

$$p_1^{f_1}p_2^{f_2} \cdots p_k^{f_k}$$

with $0 \leq f_i \leq e_i$, $i = 1, 2, . . ., k$. But these are exactly the divisors of n.

Exercise 9. Calculate $\sigma(240)$.

Both d and σ are members of an important class of number-theoretic functions: the multiplicative functions. We will now define this term, verify that d and σ are multiplicative functions, and explain why the idea is important. A function f, defined for the positive integers, is said to be *multiplicative* if and only if

$$(m, n) = 1 \qquad \text{implies} \qquad f(mn) = f(m)f(n).$$

A simple example of a multiplicative function is given by $f(n) = n$. Another is $f(n) = n^2$.

THEOREM 3. *d is multiplicative.*

PROOF. Let m and n be relatively prime. Then, no prime that divides m can divide n, and vice versa. Thus if

$$m = p_1^{e_1}p_2^{e_2} \cdots p_k^{e_k} \qquad \text{and} \qquad n = q_1^{f_1}q_2^{f_2} \cdots q_r^{f_r}$$

are the prime-power decompositions of m and n, then no q is a p and no p is a q, and the prime-power decomposition of mn is given by

$$mn = p_1^{e_1}p_2^{e_2} \cdots p_k^{e_k}q_1^{f_1}q_2^{f_2} \cdots q_r^{f_r}.$$

Applying Theorem 1, we have

$$d(mn) = ((e_1 + 1)(e_2 + 1) \cdots (e_k + 1))((f_1 + 1)(f_2 + 1) \cdots (f_r + 1))$$
$$= d(m)d(n),$$

and this proves the theorem.

THEOREM 4. *σ is multiplicative.*

PROOF. The idea is exactly the same as in Theorem 3. With m and n relatively prime, as in that theorem, we apply Theorem 2 to get

$$\sigma(mn) = \frac{p_1^{e_1+1} - 1}{p_1 - 1} \cdots \frac{p_k^{e_k+1} - 1}{d_k - 1} \cdot \frac{q_1^{f_1+1} - 1}{q_1 - 1} \cdots \frac{q_r^{f_r+1} - 1}{q_r - 1}$$
$$= \sigma(m)\sigma(n).$$

The reason multiplicative functions are important is this: if we know the value of a multiplicative function f for all prime-powers, then we can find the value of f for all positive integers. To see this, we note

THEOREM 5. *If f is a multiplicative function and the prime-power decomposition of n is $p_1^{e_1} p_2^{e_2} \cdots p_k^{e_k}$, then*

$$f(n) = f(p_1^{e_1}) f(p_2^{e_2}) \cdots f(p_k^{e_k}).$$

PROOF. The proof is by induction on k. The theorem is trivially true for $k = 1$. Suppose it is true for $k = r$. Because

$$(p_1^{e_1} p_2^{e_2} \cdots p_r^{e_r}, p_{r+1}^{e_{r+1}}) = 1,$$

we have, from the definition of a multiplicative function

$$f((p_1^{e_1} p_2^{e_2} \cdots p_r^{e_r}) p_{r+1}^{e_{r+1}}) = f(p_1^{e_1} p_2^{e_2} \cdots p_r^{e_r}) f(p_{r+1}^{e_{r+1}}).$$

From the induction assumption, the first factor is

$$f(p_1^{e_1} p_2^{e_2} \cdots p_r^{e_r}) = f(p_1^{e_1}) f(p_2^{e_2}) \cdots f(p_r^{e_r}),$$

and this, together with the preceding equation, completes the induction.

For an example, suppose that $f(p^e) = e p^{e-1}$ for all primes p and all e, $e \geq 1$. The first few values of f are

$$
\begin{array}{lcccccccccccc}
n & 2 & 3 & 4 & 5 & 6 & 7 & 8 & 9 & 10 & 11 & 12, \\
f(n) & 1 & 1 & 4 & 1 & 1 & 1 & 12 & 6 & 1 & 1 & 4, \\
\end{array}
$$

$$f(3141) = f(3^2 \cdot 349) = f(3^2) f(349) = 6 \cdot 1 = 6,$$

and we can calculate $f(n)$ for any n in a similar manner.

Exercise 10. Compute $f(n)$ for $n = 13, 14, \ldots, 24$.

We will apply Theorem 5 of this section in Section 9 to get a formula for an important number-theoretic function, Euler's ϕ-function.

Problems

1. Calculate (a) $d(42)$, (b) $d(420)$, (c) $d(4200)$.

2. Calculate (a) $\sigma(42)$, (b) $\sigma(420)$, (c) $\sigma(4200)$.

3. Use Table C (p. 218) to calculate (a) $d(10,001)$, (b) $d(10,008)$, (c) $d(100,001)$.

4. Use Table C to calculate (a) $\sigma(10,001)$, (b) $\sigma(10,008)$, (c) $\sigma(100,001)$.

5. Descartes noted, in 1638, that

$$\sigma(p^n) - p^n = \frac{p^n - 1}{p - 1}$$

for $n = 1, 2, \ldots$. Verify that this is so.

6. Cardano was the first to mention $d(n)$ when, in 1537, he said that if $p_1, p_2, \ldots, p_k$ are distinct primes, then

$$d(p_1 p_2 \cdots p_k) - 1 = 1 + 2 + 2^2 + \cdots + 2^{k-1}.$$

Verify that this is so.

7. Show that $\sigma(n)$ is odd if n is a power of two.

8. (a) Prove that if $f(n)$ is multiplicative, then so is $f(n)/n$.
   (b) Disprove that if $f(n)$ is multiplicative, then so is $f(n) - n$.

9. What is the smallest integer n such that $d(n) = 8$? Such that $d(n) = 10$?

10. Does $d(n) = k$ have a solution n for each k?

11. In 1644 Mersenne asked for a number with 60 divisors. Find one smaller than 10,000.

12. Find infinitely many n such that $d(n) = 60$.

13. Show that

$$\sum_{d|n} 1/d = \sigma(n)/n.$$

14. If p is an odd prime, for which k is $1 + p + \cdots + p^k$ odd?

15. For which n is $\sigma(n)$ odd?

16. If n is a square, show that $d(n)$ is odd.

17. If $d(n)$ is odd, show that n is a square.

18. Find all 17 solutions of

$$\frac{1}{x} + \frac{1}{y} = \frac{1}{6}$$

in integers (positive or negative).

19. How many solutions does

$$\frac{1}{x} + \frac{1}{y} = \frac{1}{N}$$

have for a given positive integer N?

20. Prove Theorem 2 by induction on k.

21. Find infinitely many n such that $\sigma(n) \le \sigma(n - 1)$.

22. If N is odd, how many solutions does $x^2 - y^2 = N$ have?

23. If N is odd, show that $x^2 - y^2 = 2N$ has no solutions.

24. Develop a formula for $\sigma_2(n)$, the sum of the squares of the positive divisors of n.

25. Guess a formula for

$$\sigma_k(n) = \sum_{d \mid n} d^k,$$

where k is a positive integer.

26. Show that the product of the positive divisors of n is $n^{d(n)/2}$.

Perfect Numbers

A number is called *perfect* if and only if it is equal to the sum of its positive divisors, excluding itself. For example, 6 is perfect, because $6 = 1 + 2 + 3$. So is 28 perfect, because $28 = 1 + 2 + 4 + 7 + 14$. But 18 is not perfect, because the sum of its positive divisors, excluding itself, is $1 + 2 + 3 + 6 + 9 = 21$. We study perfect numbers because, for mystical reasons, such numbers have been given a lot of attention in the past; because they provide practice with the σ-function; and most important, because Euler proved a satisfying theorem that allows us to determine all even perfect numbers. Long before Euler, Euclid found some perfect numbers; we will follow in his footsteps.

In symbols, the sum of the positive divisors of n, excluding itself, is $\sigma(n) - n$. Hence a number is perfect if and only if $\sigma(n) = 2n$. To find solutions to this equation, we will need to use a result proved in Section 7—namely, that σ is a multiplicative function, or

(1)   if $(m, n) = 1$,   then $\sigma(mn) = \sigma(m)\sigma(n)$.

With its aid, we can prove

THEOREM 1 (Euclid). *If $2^p - 1$ is prime, then $2^{p-1}(2^p - 1)$ is perfect.*

PROOF. Let $n = 2^{p-1}(2^p - 1)$. Because $2^p - 1$ is prime, we know that $\sigma(2^p - 1) = 2^p$. Then, noting that 2^{p-1} and $2^p - 1$ are relatively prime and applying (1), we have

$$\sigma(n) = \sigma(2^{p-1}(2^p - 1)) = \sigma(2^{p-1})\sigma(2^p - 1) = (2^p - 1) \cdot 2^p = 2n.$$

Thus n is perfect.

   As you may already have found, if you solved Problem 13, Section 2, $2^n - 1$ is composite if n is composite (because if $n = ab$, then

$$2^n - 1 = 2^{ab} - 1 = (2^a - 1)(2^{a(b-1)} + 2^{a(b-2)} + \cdots + 1)).$$

Thus $2^n - 1$ can be prime only if n is prime. In a search for perfect numbers of the form $2^{n-1}(2^n - 1)$, we thus have only to consider prime values of n. And every time we find a prime p such that $2^p - 1$ is prime, we can construct a perfect number. The first few values of $2^p - 1$ are

p	2	3	5	7	11	13
$2^p - 1$	3	7	31	127	2047	8191

and all of these except $2047 = 23 \cdot 89$ are prime. Thus we have five perfect numbers:

$$2(2^2 - 1) = 6,$$
$$2^2(2^3 - 1) = 28,$$
$$2^4(2^5 - 1) = 496,$$
$$2^6(2^7 - 1) = 8128,$$
$$2^{12}(2^{13} - 1) = 33550336.$$

An example of a larger perfect number is

$$191561942608236107294793378084303638130993721548169216.$$

   Now we will show that the numbers $2^{p-1}(2^p - 1)$ with p and $2^p - 1$ prime are the only even perfect numbers.

   THEOREM 2 (Euler). *If n is an even perfect number, then*

$$n = 2^{p-1}(2^p - 1)$$

for some prime p, and $2^p - 1$ is also prime.

   PROOF. If n is an even perfect number, then $n = 2^e m$, where m is odd and $e \geq 1$. Since $\sigma(m) > m$, we can write $\sigma(m) = m + s$, with $s > 0$. Then $2n = \sigma(n)$ becomes

$$2^{e+1}m = (2^{e+1} - 1)(m + s) = 2^{e+1}m - m + (2^{e+1} - 1)s.$$

Thus

$$m = (2^{e+1} - 1)s,$$

which says that s is a divisor of m and $s < m$. But $\sigma(m) = m + s$; thus s is the *sum* of all the divisors of m that are less than m. That is, s is the sum of a group of numbers that includes s. This is possible only if the group consists of one number alone. Therefore, the set of divisors of m smaller than m must contain only one element, and that element must be 1. That is, $s = 1$, and hence $m = 2^{e+1} - 1$ is a prime.

We repeat the argument, because it is slippery. Let the divisors of m be

$$1, d_2, d_3, \ldots, d_k, m.$$

Then $\sigma(m) = m + s$, or

$$s = 1 + d_2 + d_3 + \cdots + d_k.$$

But s is a divisor of m and $s < m$, so s *equals* one of $1, d_2, \ldots, d_k$. The only way that can be possible is if $s = 1$.

We have shown that $s = 1$. Thus $\sigma(m) = m + s = m + 1$. This says that m is prime. From (2), $m = 2^{e+1} - 1$. The only numbers of this form that can be prime are those with $e + 1$ prime. Hence $m = 2^p - 1$ for some prime p, and this completes the proof.

Thus the even perfect numbers determined in Theorem 1 are the only even perfect numbers. As for odd perfect numbers, no one knows if there are any, and no one has proved that none can exist. It is known that if there is an odd perfect number, then it is quite large: in 1967 it was announced that it must be greater than

$$1,000,000,000,000,000,000,000,000,000,000,000,$$

and there are many other conditions that odd perfect numbers must satisfy. But no combination of conditions has so far served to show that there are no odd perfect numbers: it may be that there is one, but so huge that it is out of the range of human computation.

The problem of finding even perfect numbers is, after Theorem 2, the same as the problem of determining primes p such that $2^p - 1$ is also prime. Primes of the form $2^p - 1$ are called *Mersenne primes*. In the seventeenth century, Mersenne claimed that $2^p - 1$ was prime for

$$p = 2, 3, 5, 7, 13, 17, 31, 67, 127, 257,$$

and for no other primes less than 257. His guess was not accurate: he erred in including 67 and 257, for

$$2^{67} - 1 = 193707721 \cdot 761838257287,$$

and $2^{257} - 1$ is also composite. He further erred in excluding 19, 61, 89, and 107. But don't think harshly of Mersenne: in the seventeenth century, there were no mathematical journals to announce new discoveries; instead, almost everyone wrote to Mersenne, and Mersenne wrote to almost everyone else, enclosing the latest mathematical news. He thus spread the results of Fermat, for one, and sped the development of mathematics; it is fitting that he have a set of primes named after him. The complete list of currently known primes p such that $2^p - 1$ is prime is

$$2, 3, 5, 7, 13, 17, 19, 31, 61, 89, 107, 127, 521, 607, 1279,$$
$$2203, 2281, 3217, 4253, 4423, 9689, 9941, 11213,$$

and to each of these there corresponds an even perfect number: 23 in all. The first twelve were discovered before the invention of high-speed computers; the later ones are so enormous as to be beyond the reach of hand computation. The search for Mersenne primes has gone on (the last few were discovered very recently) in the hope of seeing some sort of pattern in the primes p, so that theorems could be guessed and maybe proved. Some conjectures have been advanced, but without any indication as to how to go about proving them, and in fact, no important theorems have been proved. It is not even known if there are infinitely many such primes.

To close the section, we will mention another kind of number that used to excite some people: *amicable numbers*. Consider 220 and 284. Since $220 = 2^2 \cdot 5 \cdot 11$, it follows that

$$\sigma(220) - 220 = \sigma(2^2)\sigma(5)\sigma(11) - 220 = 7 \cdot 6 \cdot 12 - 220$$
$$= 504 - 220 = 284.$$

And since $284 = 2^2 \cdot 71$, we have

$$\sigma(284) - 284 = \sigma(2^2)\sigma(71) - 284 = 7 \cdot 72 - 284$$
$$= 504 - 284 = 220.$$

So, in some sense, 220 and 284 go together. In general, we say that m and n are *amicable* (or are an *amicable pair*) if and only if

$$\sigma(m) - m = n \qquad \text{and} \qquad \sigma(n) - n = m.$$

Equivalently, we could say that m and n are amicable if and only if

$$\sigma(m) = \sigma(n) = m + n.$$

Exercise 1. Verify that 1184 and 1210 are amicable.

It used to be thought by some that if one person carried a talisman of some sort containing the number 220, and another person had one with 284, they would be favorably disposed to each other. Numbers undeniably have power: it might be worth a try today. Amicable numbers attracted the attention of respectable mathematicians as late as the early twentieth century (the amicable pair in Exercise 1 was first discovered as late as 1866); Euler found many such pairs, and long lists of them exist. (Besides those already mentioned, the amicable pairs less than 10,000 are 2620, 2924; 5020, 5564;

and 6232, 6368.) But there are no general theorems on amicable numbers as beautiful as Euclid's and Euler's theorems on perfect numbers, and amicable numbers are not taken very seriously today. We will leave them—and other numbers of similar type—to the problems that follow.

Problems

1. Verify that $17296 = 2^4 \cdot 23 \cdot 47$ and $18416 = 2^4 \cdot 1151$ are amicable. (This pair, discovered by Fermat, was the first to be found after 220, 284, which was known to the Greeks.)

2. It was long thought that even perfect numbers end alternately in 6 and 8. Show that this is wrong by verifying that
$$2^{12}(2^{13} - 1) \equiv 2^{16}(2^{17} - 1) \equiv 6 \ (\mathrm{mod}\ 10).$$

3. In 1575, it was observed that every even perfect number is a triangular number. Show that this is so.

4. In 1652, it was observed that
$$6 = 1 + 2 + 3,$$
$$28 = 1 + 2 + 3 + 4 + 5 + 6 + 7,$$
$$496 = 1 + 2 + 3 + \cdots + 31.$$

Can this go on?

5. Show that if m and n are amicable, then
$$\sum_{d|m} d = \sum_{d|n} d = m + n.$$

6. Show that if m and n are amicable, then
$$\left(\sum_{d|m} 1/d\right)^{-1} + \left(\sum_{d|n} 1/d\right)^{-1} = 1.$$

7. Let
$$p = 3 \cdot 2^e - 1,$$
$$q = 3 \cdot 2^{e-1} - 1,$$
$$r = 3^2 \cdot 2^{2e-1} - 1,$$

where e is a positive integer. If p, q, and r are all prime, show that $2^e pq$ and $2^e r$ are amicable. ($e = 2$, 4, and 7 give amicable pairs, but for no other $e \leq 200$ are p, q, and r all prime.)

8. Show that no prime can be one of an amicable pair.

9. If p^e is one of an amicable pair, show that
$$\sigma(p^e) = \sigma\left(\frac{p^e - 1}{p - 1}\right).$$

10. (a) Show that $\sigma(1 + p) < 1 + p + p^2$.
    (b) Use this to show that p^2 can never be one of an amicable pair.

11. If $\sigma(n) = kn$, then n is called a *k-perfect number*. Verify that 672 is 3-perfect and $2,178,540 = 2^2 \cdot 3^2 \cdot 5 \cdot 7^2 \cdot 13 \cdot 19$ is 4-perfect.

12. Let $s(n) = \sigma(n) - 2n$. If $s(n) = 0$, then n is perfect. If $s(n) > 0$, we say that n is *abundant*. If $s(n) < 0$, we say that n is *deficient*.

(a) Verify that 12 and 24 are abundant and that 8 and 14 are deficient.
(b) Classify the positive integers less than or equal to 20 as abundant, deficient, or perfect.
(c) It was long thought that every abundant number was even. Show that 945 is abundant.
(d) Show that $n(n + 1)$ is abundant for $n = 3, 4, 5, 6, 7, 8, 9$, but deficient for $n = 10$.
(e) If $p > 3$ and $2p + 1$ are prime, show that $2p(2p + 1)$ is deficient: in fact, $s(2p(2p + 1)) = -2p^2 + 8p + 6$.
(f) Show that pq ($pq \neq 6$) is deficient.
(g) Let $n = 2^k(2^{k+1} - 1)$. If $2^{k+1} - 1$ is composite, show that n is abundant.
(h) If $n = 2^k(2^{k+1} - 1)$ is perfect, $d \mid n$, and $1 < d < n$, show that d is deficient.
(i) If $n = 2^k(2^{k+1} - 1)$ is perfect and $p < 2^{k+1} - 1$ is prime, show that $2^k p$ is abundant.

13. Show that p^e (p a prime, greater than 2 and $e \geq 1$) is deficient.

14. If a has order e modulo p and $a \neq 1$, show that all even perfect numbers end in 6 or 8.

15. If n is an even perfect number and $n > 6$, show that $n \equiv 1 \pmod 9$.

16. Show that if p is odd, then

$$2^{p-1}(2^p - 1) \equiv 1 + 9 \binom{p}{2} \pmod{81}.$$

$\left(\text{For the notation } \dbinom{m}{n}, \text{ see Appendix B, p. 196.}\right)$

17. Euler showed that any odd perfect number must be of the form

$$p^{4a+1}Q^2,$$

p an odd prime, a and Q integers. Fill in any missing details in this sketch of his proof:

Let $n = P_1 P_2 \cdots P_k$ be the decomposition of n into powers of distinct odd primes. Let $Q_i = \sigma(P_i)$, $i = 1, 2, \ldots, k$. If $\sigma(n) = 2n$, then

$$2P_1 P_2 \cdots P_k = Q_1 Q_2 \cdots Q_k.$$

Thus, one of $Q_1, Q_2, \ldots, Q_k$—say Q_1—is double an odd number, and the remaining ones are odd. Thus $P_2, P_3, \ldots P_k$ are even powers of primes. Also, $P_1 = p^{4a+1}$ for some prime p and integer a.

18. Here is Euler's original proof of Theorem 2. Fill in any missing details
Let $n = 2^k m$ be perfect, m odd. The sum, $(2^{k+1} - 1)\sigma(m)$, of the divisors of n must equal $2n$. Thus

$$m/\sigma(m) = (2^{k+1} - 1)/2^{k+1},$$

a fraction in lowest terms. Hence $m = (2^{k+1} - 1)c$ for some integer c. If

$c = 1$, then $m = 2^{k+1} - 1$ must be prime, because $\sigma(m) = 2^{k+1}$. If $c > 1$, then $\sigma(m) \geq m + (2^{k+1} - 1) + c + 1$. Thus

$$\frac{\sigma(m)}{m} \geq \frac{2^{k+1}(c + 1)}{m} > \frac{2^{k+1}}{2^{k+1} - 1},$$

a contradiction.

SECTION 9

Euler's Theorem and Function

Fermat's Theorem states that

$$\text{if} \quad (a, p) = 1, \quad \text{then} \quad a^{p-1} \equiv 1 \pmod{p}.$$

It is natural to ask if there is a generalization of this to composite moduli: Given any integer m, is there a number $f(m)$ such that $a^{f(m)} \equiv 1 \pmod{m}$? We note that this cannot hold unless $(a, m) = 1$, for if a and m have a common divisor greater than 1, then $m \mid (a^k - 1)$ is impossible for any $k > 0$. Let us look at tables of powers of $a \pmod{m}$, where a and m are relatively prime, for $m = 6$, 9, and 10:

$m = 6$

a	a^2
1	1
5	1

$m = 9$

a	a^2	a^3	a^4	a^5	a^6
1	1	1	1	1	1
2	4	8	7	5	1
4	7	1	4	7	1
5	7	8	4	2	1
7	4	1	7	4	1
8	1	8	1	8	1

$m = 10$

a	a^2	a^3	a^4
1	1	1	1
3	9	7	1
7	9	3	1
9	1	9	1

Evidently,

$$a^2 \equiv 1 \pmod{6} \quad \text{if} \quad (a, 6) = 1,$$
$$a^6 \equiv 1 \pmod{9} \quad \text{if} \quad (a, 9) = 1,$$
$$a^4 \equiv 1 \pmod{10} \quad \text{if} \quad (a, 10) = 1,$$

so the number $f(m)$ exists for $m = 6$, 9, and 10.

Exercise 1. Show that $a^6 \equiv 1 \pmod{14}$ for all a relatively prime to 14.

If your eye is very sharp indeed, you might have noticed that $f(6) = 2$ and that there are two positive integers less than 6 and relatively prime to 6; $f(9) = 6$, and there are six positive integers less than 9 and relatively prime to it; $f(10) = 4$, and there are four positive integers less than 10 and relatively prime to it; and a similar statement holds for 14. So, you might guess (and if you worked more examples, you would almost certainly guess)

THEOREM 1. *Suppose that $m \geq 2$ and $(a, m) = 1$. If $\phi(m)$ denotes the number of positive integers less than m and relatively prime to it, then*

$$a^{\phi(m)} \equiv 1 \ (mod \ m).$$

Note that this guess is correct in the special case when $m = p$, a prime. Every positive integer less than p is relatively prime to it, so $\phi(p) = p - 1$, and we know that $a^{p-1} \equiv 1 \pmod{p}$ when $(a, p) = 1$.

In this section we will prove Theorem 1, and we will develop a formula for calculating $\phi(n)$ from the prime-power decomposition of n. Theorem 1 was first proved by Euler, and ϕ is called *Euler's ϕ-function.*

The idea used to prove Fermat's Theorem is that

if $(a, p) = 1$, then the least residues (mod p) of

$a, 2a, \ldots, (p - 1)a$ are a permutation of

$1, 2, \ldots, p - 1.$

This is also the key to Euler's generalization:

LEMMA 1. *If $(a, m) = 1$ and $r_1, r_2, \ldots, r_{\phi(m)}$ are the positive integers less than m and relatively prime to m, then the least residues (mod m) of*

(1) $ar_1, ar_2, \ldots, ar_{\phi(m)}$

are a permutation of

$$r_1, r_2, \ldots, r_{\phi(m)}.$$

Exercise 2. Verify that Lemma 1 is true if $m = 10$ and $a = 3$.

PROOF OF LEMMA 1. Since there are exactly $\phi(m)$ numbers in the set (1), to prove that their least residues are a permutation of the $\phi(m)$ numbers $r_1, r_2, \ldots, r_{\phi(m)}$ we have to show that they are all different and that they are all relatively prime to m. To show that they are all different, suppose that

$$ar_i \equiv ar_j \pmod{m}$$

for some i and j ($1 \leq i \leq \phi(m)$, $1 \leq j \leq \phi(m)$). Since $(a, m) = 1$, we can cancel a from both sides of the congruence to get $r_i \equiv r_j \pmod{m}$. Since r_i and r_j are least residues (mod m), it follows that $r_i = r_j$. Hence, $r_i \neq r_j$ implies $ar_i \not\equiv ar_j \pmod{m}$, and so the numbers in (1) are all different.

To prove that all the numbers in (1) are relatively prime to m, suppose that p is a prime common divisor of ar_i and m for some i, $1 \leq i \leq \phi(m)$. Since p is prime, either $p \mid a$ or $p \mid r_i$. Thus either p is a common divisor of a and m or of r_i and m. But $(a, m) = (r_i, m) = 1$, so both cases are impossible. Hence $(ar_i, m) = 1$ for each i, $i = 1, 2, \ldots, \phi(m)$.

The proof of Euler's Theorem proceeds similarly to the proof of Fermat's Theorem:

PROOF OF THEOREM 1. From Lemma 1 we know that

$$r_1 r_2 \cdots r_{\phi(m)} \equiv (ar_1)(ar_2) \cdots (ar_{\phi(m)})$$
$$\equiv a^{\phi(m)}(r_1 r_2 \cdots r_{\phi(m)}) \pmod{m}.$$

Since each of $r_1, r_2, \ldots, r_{\phi(m)}$ is relatively prime to m, it follows that their product is also; thus that factor may be cancelled in the last congruence, and we get

$$1 \equiv a^{\phi(m)} \pmod{m}.$$

The rest of the section will be mainly devoted to the properties of ϕ; our goal is to find a way of calculating $\phi(n)$ by some method other than actually counting all the positive integers less than n and relatively prime to it.

Exercise 3. Verify that the entries in the following table are correct.

n	2	3	4	5	6	7	8	9	10
$\phi(n)$	1	2	2	4	2	6	4	6	4

In order that $\phi(n)$ exist for all positive integers n, we *define* $\phi(1)$ to be 1. With this definition, Theorem 1 is true for all positive integers m.

Exercise 4. Verify that $3^{\phi(8)} \equiv 1 \pmod{8}$.

Exercise 5. Which positive integers are less than 4 and relatively prime to it? What is the answer if 4 is replaced by 8? By 16? Can you induce a formula for $\phi(2^n)$, $n = 1, 2, \ldots$?

In general, it is not hard to see what $\phi(p^n)$ is, where p is a prime and n is a positive integer.

LEMMA 2. $\phi(p^n) = p^{n-1}(p - 1)$ *for all positive integers n.*

PROOF. The positive integers less than or equal to p^n which are *not* relatively prime to p^n are exactly the multiples of p:

$$1 \cdot p, \, 2 \cdot p, \, 3 \cdot p, \, \ldots, \, (p^{n-1})p,$$

and there are p^{n-1} of them. Since there are in all p^n positive integers less than or equal to p^n, we have

$$\phi(p^n) = p^n - p^{n-1} = p^{n-1}(p - 1).$$

Exercise 6. Verify that the formula is correct for $p = n = 3$.

Thus we know ϕ for all prime-powers. If we knew that ϕ was a multiplicative function, then we could apply Theorem 5 of Section 7 to get a formula for $\phi(n)$. That ϕ is in fact multiplicative we will now demonstrate in a theorem whose proof, in common with many other proofs in number theory, is neither long, technical, nor complicated; it is just hard. First we need an easy lemma:

LEMMA 3. *If* $(a, m) = 1$ *and* $a \equiv b \pmod{m}$, *then* $(b, m) = 1$.

PROOF. This follows from the fact that $b = a + km$ for some k.

COROLLARY. If the least residues (mod m) of

(2) $r_1, r_2, \ldots, r_m$

are a permutation of $0, 1, \ldots, m - 1$, then (2) contains exactly $\phi(m)$ elements relatively prime to m.

We can now prove

THEOREM 2. ϕ *is multiplicative*.

PROOF. Write the numbers from 1 to mn as follows:

$$\begin{array}{ccccc}
1 & m+1 & 2m+1 & \ldots & (n-1)m+1 \\
2 & m+2 & 2m+2 & \ldots & (n-1)m+2 \\
& & \cdots & & \\
m & 2m & 3m & \ldots & mn
\end{array}$$

Suppose that $(m, r) = d$ and $d > 1$. Then we claim that no element in the rth row of the array:

$$r \quad m+r \quad 2m+r \quad \ldots \quad km+r \quad \ldots \quad (n-1)m+r$$

is relatively prime to mn. This is so because if $d \mid m$ and $d \mid r$, then $d \mid mn$ and $d \mid (km + r)$ for any k. So, if we are looking for numbers that are relatively prime to mn, we will not find any except in those rows whose first element is relatively prime to m.

Exercise 7. How many such rows are there?

For an example, let us take $n = 5$ and $m = 6$. Then the array is

$$
\begin{array}{ccccc}
1 & 7 & 13 & 19 & 25 \\
2 & 8 & 14 & 20 & 26 \\
3 & 9 & 15 & 21 & 27 \\
4 & 10 & 16 & 22 & 28 \\
5 & 11 & 17 & 23 & 29 \\
6 & 12 & 18 & 24 & 30
\end{array}
$$

No element in the second, third, fourth, or sixth row is relatively prime to $mn = 30$, because the first element in each of those rows is not relatively prime to $m = 6$. All numbers relatively prime to 30 are found in the two remaining rows,

$$
\begin{array}{ccccc}
1 & 7 & 13 & 19 & 25 \\
5 & 11 & 17 & 23 & 29
\end{array}
$$

Suppose we can show that there are exactly $\phi(n)$ numbers relatively prime to mn in each of the rows that have first elements relatively prime to m. Since there are $\phi(m)$ such rows, it will follow that the number of integers in the whole array that are relatively prime to mn is $\phi(n)\phi(m)$: that is, $(n, m) = 1$, and $\phi(mn) = \phi(m)\phi(n)$, and the theorem will be proved. But the numbers in the rth row (where r and m are relatively prime) are

(3) $$r, m + r, 2m + r, \ldots, (n - 1)m + r,$$

and we claim that their least residues (mod n) are a permutation of

(4) $$0, 1, 2, \ldots, (n - 1).$$

To verify this claim, all we have to do is show that no two of the numbers in (3) are congruent (mod n), because (3) contains n elements, just as does (4). This is easy: suppose that

$$km + r \equiv jm + r \ (\text{mod } n),$$

with $0 \leq k < n$ and $0 \leq j < n$. Then $km \equiv jm \,(\text{mod } n)$, and since $(m, n) = 1$, we have $k \equiv j \,(\text{mod } n)$. On account of the inequalities on k and j, it follows that $k = j$. Hence, if $k \neq j$, then $km + r \not\equiv jm + r \,(\text{mod } n)$, and no two elements of (3) are congruent (mod n).

By the Corollary to Lemma 3, we have that (3) contains exactly $\phi(n)$ elements relatively prime to n. But from Lemma 3, every element in the rth row of the array is relatively prime to m. It follows that the rth row of the array contains exactly $\phi(n)$ elements relatively prime to mn. As we noted before, this is enough to complete the proof.

In the example, the least residues (mod 5) of the numbers in the two rows that included all of the elements relatively prime to 30 are

$$1 \quad 2 \quad 3 \quad 4 \quad 0$$
$$0 \quad 1 \quad 2 \quad 3 \quad 4$$

and each row contains $\phi(5) = 4$ numbers relatively prime to 30. Thus $8 = \phi(30) = \phi(6)\phi(5)$.

We can now get a formula for $\phi(n)$:

THEOREM 3. *If n has a prime-power decomposition given by*

$$n = p_1^{e_1} p_2^{e_2} \cdots p_k^{e_k},$$

then

$$\phi(n) = p_1^{e_1-1}(p_1 - 1)p_2^{e_2-1}(p_2 - 1) \cdots p_k^{e_k-1}(p_k - 1).$$

PROOF. Because ϕ is multiplicative, Theorem 5 of Section 7 applies to give

$$\phi(n) = \phi(p_1^{e_1})\phi(p_2^{e_2}) \cdots \phi(p_k^{e_k}).$$

If we apply Lemma 2 to each term on the right, the theorem is proved.

Exercise 8. Use Theorem 3 to calculate $\phi(72)$, $\phi(74)$, and $\phi(76)$.

The formula of Theorem 3 can be written in another form, which is neater and sometimes useful:

COROLLARY. If $n = p_1^{e_1} p_2^{e_2} \cdots p_k^{e_k}$, then

$$\phi(n) = n\left(1 - \frac{1}{p_1}\right)\left(1 - \frac{1}{p_2}\right) \cdots \left(1 - \frac{1}{p_k}\right).$$

The proof of this corollary is left to the reader.

We conclude this section with a theorem we will need in the next section.

Exercise 9. Calculate $\displaystyle\sum_{d|n} \phi(d)$

(a) For $n = 12, 13, 14, 15,$ and 16.
(b) For $n = 2^k, k \geq 1$.
(c) For $n = p^k, k \geq 1$ and p an odd prime.

You should have guessed by now that the following theorem is true.

THEOREM 4. *If $n \geq 1$, then*

$$\sum_{d|n} \phi(d) = n.$$

PROOF. It would be natural to try to apply the formula of Theorem 3 to get this result. This would be difficult; instead we use a clever idea first thought of by Gauss. Consider the integers $1, 2, \ldots, n$. We will put one of these integers in class C_d if and only if its greatest common divisor with n is d. For example, if $n = 12$, we have

$$
\begin{aligned}
C_1 &= \{1, 5, 7, 11\}, & C_2 &= \{2, 10\}, \\
C_3 &= \{3, 9\}, & C_4 &= \{4, 8\}, \\
C_6 &= \{6\}, & C_{12} &= \{12\}.
\end{aligned}
$$

Exercise 10. What are the classes C_d for $n = 14$?

We have m in C_d if and only if $(m, n) = d$; that is, if and only if $(m/d, n/d) = 1$. Thus, from the definition of the Euler ϕ-function, the number of elements in class C_d is $\phi(n/d)$.

Exercise 11. Check that this is correct for $n = 12$ and $n = 14$.

Since there is a class for each divisor of n, the total number of elements in all the classes C_d is

$$
\sum_{d \mid n} \phi(n/d).
$$

But this is just n, since each integer $1, 2, \ldots, n$ is in exactly one class. Hence

$$
n = \sum_{d \mid n} \phi(n/d) = \sum_{d \mid n} \phi(d),
$$

and the theorem is proved.

Problems

1. Calculate $\phi(42)$, $\phi(420)$, and $\phi(4200)$.

2. Calculate $\phi(10{,}001)$ and $\phi(100{,}001)$.

3. Verify that $a^8 \equiv 1 \pmod{15}$ if $(a, 15) = 1$.

4. Which are the positive integers less than 18 and prime to it? Verify that Lemma 1 is true with $m = 18$ and $a = 5$.

5. Perfect numbers satisfy $\sigma(n) = 2n$. Which n satisfy $\phi(n) = 2n$?

6. Show that if n is odd, then $\phi(4n) = 2\phi(n)$.

7. $1 + 2 = (3/2)\phi(3)$, $\quad 1 + 3 = (4/2)\phi(4)$, $\quad 1 + 2 + 3 + 4 = (5/2)\phi(5)$, $1 + 5 = (6/2)\phi(6)$, $1 + 2 + 3 + 4 + 5 + 6 = (7/2)\phi(7)$, and $1 + 3 + 5 + 7 = (8/2)\phi(8)$. Guess a theorem.

8. Show that

$$\sum_{p \leq x} \sigma(p) - \sum_{p \leq x} \phi(p) = 2 \sum_{p \leq x} d(p).$$

9. Prove Lemma 3 by starting with the fact that there are integers r and s such that $ar + ms = 1$.

10. If $(a, m) = 1$, show that any x such that

$$x \equiv ca^{\phi(m)-1} \pmod{m}$$

satisfies $ax \equiv c \pmod{m}$.

11. (a) If p is an odd prime, how many elements in the sequence

$$1 \cdot 2, \, 2 \cdot 3, \, 3 \cdot 4, \, \ldots, \, p(p+1)$$

are relatively prime to p?
(b) If p is an odd prime, how many elements in the sequence

$$1 \cdot 2, \, 2 \cdot 3, \, 3 \cdot 4, \, \ldots, \, p^2(p^2 + 1)$$

are relatively prime to p?
(c) Any guesses for a general theorem?

12. Find four solutions of $\phi(n) = 16$.

13. Let $n = dm$. Show that there are $\phi(m)$ positive integers less than n whose greatest common divisor with n is d.

14. Show that $\phi(mn) > \phi(m)\phi(n)$ if m and n have a common factor greater than 1.

15. Let $\phi^{(2)}(n) = \phi(\phi(n))$, $\phi^{(3)}(n) = \phi(\phi^{(2)}(n))$, and so on. Let $e(n)$ denote the smallest integer such that

$$\phi^{(e(n))}(n) = 2.$$

Calculate $e(n)$ for

(a) $n = 3, 4, 5, 6, 7, 8, 9$.
(b) $n = 2^k, k \geq 2$.
(c) $n = 3^k, k \geq 1$.
(d) $n = 2^k 3^j, k \geq 1, j \geq 1$.

16. Show that $(m, n) = 2$ implies $\phi(mn) = 2\phi(m)\phi(n)$.

17. If $(m, n) = p$, how is $\phi(mn)$ related to $\phi(m)\phi(n)$?

18. Show that $\phi(n) = n/2$ if and only if $n = 2^k$ for some positive integer k.

19. Show that $\phi(n) = n/3$ if and only if $n = 2^k 3^j$ for some positive integers k and j.

20. Show that if $6 \mid n$, then $\phi(n) \leq n/3$.

21. Show that if $n - 1$ and $n + 1$ are both primes and $n > 4$, then $\phi(n) \leq n/3$.

22. Suppose we know that

$$\phi(np) = p\phi(n) \quad \text{if} \quad p \mid n, \text{ and}$$
$$\phi(np) = (p - 1)\phi(n) \quad \text{if} \quad p \nmid n.$$

Deduce from these the formula for $\phi(n)$.

23. Calculate

$$\sum_{d|n} (-1)^{n/d} \phi(d)$$

for

(a) $n = 12, 13, 14, 15, 16$.
(b) $n = p$, p an odd prime.
(c) $n = 2^k$, $k \geq 1$.
(d) $n = p^k$, $k \geq 1$ and p an odd prime.
(e) Guess a theorem.

24. Find all n such that $4 \nmid \phi(n)$.

25. Show that $\phi(n) = 14$ is impossible.

26. Find a positive integer k, $k \neq 7$, such that $\phi(n) = 2k$ is impossible.

27. Prove the theorem of Problem 7.

28. In a rectangular coordinate system, put a dot at the point (n, m) if and only if n and m are relatively prime. Consider one-by-one boxes.

(a) Can any such box have all four corners dotted?
(b) Find a box with no corners dotted.
(c) Let p be an odd prime. In the row of boxes

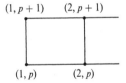
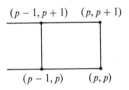

how many have three corners dotted?

Primitive Roots and Indices

In Theorem 1 of the last section, we saw that if $(a, m) = 1$, then there is a positive integer t such that $a^t \equiv 1 \pmod{m}$, namely $t = \phi(m)$. This can be proved independently of that theorem as follows. If $(a, m) = 1$, then the least residues (mod m) of a, a^2, a^3, ... are all relatively prime to m. There are $\phi(m)$ least residues (mod m) that are relatively prime to m and infinitely many powers of a: it follows that there are positive integers j and k such that $j \neq k$ and $a^j \equiv a^k \pmod{m}$. Since $(a, m) = 1$, the smaller power of a in the last congruence may be canceled, and we have either

$$a^{j-k} \equiv 1 \pmod{m} \qquad \text{or} \qquad a^{k-j} \equiv 1 \pmod{m}.$$

So, if $(a, m) = 1$, then there is a positive integer t such that $a^t \equiv 1 \pmod{m}$. In fact, there are infinitely many, since it follows from $a^{\phi(m)} \equiv 1 \pmod{m}$ for $(a, m) = 1$ that, for any positive integer k,

$$a^{t+k\phi(m)} \equiv a^t(a^k)^{\phi(m)} \equiv a^t \equiv 1 \pmod{m}.$$

The *smallest* such positive integer will be called the *order* of a modulo m. For example, modulo 7 we have

a	a^2	a^3	a^4	a^5	a^6
1	1	1	1	1	1
2	4	1	2	4	1
3	2	6	4	5	1
4	2	1	4	2	1
5	4	6	2	3	1
6	1	6	1	6	1

so 3 and 5 have order six (mod 7), 2 and 4 have order three, 6 has order two, and 1 has order one. When this idea was introduced by Gauss, he called *t the exponent to which a belongs* (mod *m*). Gauss's long phrase is more commonly used than our shorter one.

Exercise 1. What are the orders of 3, 5, and 7, modulo 8?

The order of an integer (mod *m*) cannot be just any number: as we now show, it must be a divisor of $\phi(m)$.

LEMMA 1. *If* $(a, m) = 1$, $a^n \equiv 1$ *(mod m) with* $n > 0$, *and a has order t (mod m), then* $t \mid n$.

PROOF. Suppose that $(t, n) = d$. Then we know that there are integers *r* and *s* such that $rt + sn = d$. Hence

$$a^d = a^{rt+sn} = (a^t)^r(a^n)^s.$$

But by assumption, $a^n \equiv a^t \equiv 1$ (mod *m*); it follows that $a^d \equiv 1$ (mod *m*). But *t* was assumed to be the smallest positive integer such that $a^t \equiv 1$ (mod *m*). Hence $t \leq d$. But since $(t, n) = d$, then *d* is a divisor of *t*. and $d \leq t$. Thus $d = t$, and it follows that *t* is a divisor of *n*. The proof neglects the fact that one of *r* and *s* is negative. It can be repaired by defining a^{-n} (mod *m*) for $n > 0$ by $a^n (a^{-n}) \equiv 1$ (mod *m*).

THEOREM 1. *If* $(a, m) = 1$ *and a has order t (mod m), then* $t \mid \phi(m)$.

PROOF. By Euler's extension of Fermat's Theorem (Theorem 1 of the last section), $a^{\phi(m)} \equiv 1$ (mod *m*). The result then follows from Lemma 1 on setting $n = \phi(m)$.

Exercise 2. What order can an integer have (mod 9)? Find an example of each.

As an example of the application of this idea, we prove

THEOREM 2. *If p and q are odd primes and* $q \mid a^p - 1$, *then either* $q \mid a - 1$ *or* $q = 2kp + 1$ *for some integer k.*

PROOF. Since $q \mid a^p - 1$, we have $a^p \equiv 1$ (mod *q*). So, by Lemma 1, the order of *a* (mod *q*) is a divisor of *p*. That is, *a* has order 1 or *p*. If the order of *a* is 1, then $a^1 \equiv 1$ (mod *q*), so $q \mid a - 1$. If on the other hand the order of *a* is *p*, then by Theorem 1, $p \mid \phi(q)$, or $p \mid q - 1$. That is, $q - 1 = rp$ for some integer *r*. Since *p* and *q* are odd, *r* must be even, and this completes the proof.

COROLLARY. Any divisor of $2^p - 1$ is of the form $2kp + 1$.

Exercise 3. Using the corollary, what is the smallest prime divisor of $2^{19} - 1$?

If the order of $a \pmod m$ is $\phi(m)$, then we will say that a is a *primitive root* of m. Primitive roots, and numbers that have them, are of special interest because of the following property:

THEOREM 3. *If g is a primitive root of m, then the least residues, modulo m, of*

$$g, g^2, \ldots, g^{\phi(m)}$$

are a permutation of the $\phi(m)$ positive integers less than m and relatively prime to it.

For example, 2 is a primitive root of 9, and the powers

$$2, 2^2, 2^3, 2^4, 2^5, 2^6$$

are, $\pmod 9$,

$$2, 4, 8, 7, 5, 1.$$

PROOF OF THEOREM 3. Since $(g, m) = 1$, each power of g is relatively prime to m. Moreover, if

$$g^j \equiv g^k \pmod m, \qquad 1 \le j \le \phi(m), \quad 1 \le k \le \phi(m),$$

and if we assume that $j \ge k$ (as we may do with no loss of generality), then

$$g^{j-k} \equiv 1 \pmod m.$$

But $0 \le j - k < \phi(m)$; because g is a primitive root of m,

$$g^t \equiv 1 \pmod m, \qquad 0 \le t < \phi(m)$$

is possible only for $t = 0$. Hence $j - k = 0$ and $j = k$.

Exercise 4. Show that 3 is a primitive root of 7.

Exercise 5. Find, by trial, a primitive root of 10.

Not every integer has primitive roots—for example, 8 does not, as we saw in Exercise 1. We will now set out to show that each prime has a primitive root. The proof is not easy, requires a good deal of preparation (Lemmas 2 to 4), and, because it is an existence proof, it does not show how to find the primitive root. For these reasons, you do not lose too much if you take the result on faith; be assured that it is true.

If a has order $d \pmod p$, then $(a^k)^d \equiv 1 \pmod p$ for $k = 1, 2, \ldots, d$. Some of the integers $a^2, a^3, \ldots, a^d$ have order d, and some have smaller order. For example, a^d has order 1.

Exercise 6. The order of 3 $\pmod{10}$ is 4. Find the orders of 3^2, 3^3, and $3^4 \pmod{10}$.

The next lemma tells which powers of a have the same order as a.

LEMMA 2. *Suppose that a has order $d \pmod p$. Then a^k has order $d \pmod p$ if and only if $(k, d) = 1$.*

PROOF. Suppose that $(k, d) = 1$, and denote the order of a^k by t. We have

$$1 \equiv (a^d)^k \equiv (a^k)^d \pmod p,$$

so from Lemma 1, $t \mid d$. Because t is the order of a^k,

$$(a^k)^t \equiv a^{kt} \equiv 1 \pmod p,$$

so from Lemma 1 again, $d \mid kt$. Since $(k, d) = 1$, it follows that $d \mid t$. This fact, together with the fact that $t \mid d$, implies $t = d$.

To prove the converse, suppose that a and a^k have order d and that $(k, d) = r$. Then

$$1 \equiv a^d \equiv (a^d)^{k/r} \equiv (a^k)^{d/r} \pmod p;$$

because d is the order of a^k, Lemma 1 says that $d \mid (d/r)$. This implies that $r = 1$.

We now need a lemma about the solutions of polynomial congruences $\pmod p$. Though it is an important theorem, we do not use it elsewhere.

LEMMA 3. *If f is a polynomial of degree n, then*

(1) $$f(x) \equiv 0 \pmod p$$

has at most n solutions.

PROOF. Let

$$f(x) = a_n x^n + a_{n-1} x^{n-1} + \cdots + a_0$$

have degree n; that is, $a_n \not\equiv 0 \pmod p$. We prove the lemma by induction. For $n = 1$,

$$a_1 x + a_0 \equiv 0 \pmod p$$

has but one solution, since $(a_1, p) = 1$. Suppose that the lemma is true for polynomials of degree $n - 1$, and suppose that f has degree n. Either $f(x) \equiv 0 \pmod p$ has no solutions or it has at least one. In the first case, the lemma is true. In the second case, suppose that r is a solution. That is, $f(r) \equiv 0 \pmod p$, and r is a least residue $\pmod p$. Then because $x - r$ is a factor of $x^t - r^t$ for $t = 0, 1, \ldots, n$, we have

$$\begin{aligned} f(x) &\equiv f(x) - f(r) \\ &\equiv a_n(x^n - r^n) + a_{n-1}(x^{n-1} - r^{n-1}) + \cdots + a_1(x - r) \\ &\equiv (x - r)g(x) \pmod p, \end{aligned}$$

where g is a polynomial of degree $n - 1$. Suppose that s is also a solution of (1). Thus

$$f(s) \equiv (s - r)g(s) \equiv 0 \ (\text{mod } p).$$

Because p is a prime it follows that

$$s \equiv r \ (\text{mod } p) \qquad \text{or} \qquad g(s) \equiv 0 \ (\text{mod } p);$$

from the induction assumption, the second congruence has at most $n - 1$ solutions. Since the first congruence has just one solution, the proof is complete.

Note that Lemma 3 is not true if the modulus is not a prime. For example,

$$x^2 + x \equiv 0 \ (\text{mod } 6)$$

has solutions 0, 2, 3, and 5.

LEMMA 4. *If $d \mid p - 1$, then $x^d \equiv 1$ (mod p) has exactly d solutions.*

PROOF. From Fermat's Theorem, the congruence $x^{p-1} \equiv 1$ (mod p) has exactly $p - 1$ solutions, namely $1, 2, \ldots, p - 1$. Moreover,

$$x^{p-1} - 1 = (x^d - 1)(x^{p-1-d} + x^{p-1-2d} + \cdots + 1)$$
$$= (x^d - 1)h(x).$$

From Lemma 3, we know that $h(x) \equiv 0$ (mod p) has at most $p - 1 - d$ solutions. Hence $x^d \equiv 1$ (mod p) has at least d solutions. Applying Lemma 3 again, we see that it has exactly d solutions.

We are at last prepared to prove

THEOREM 4. *Every prime p has $\phi(p - 1)$ primitive roots.*

PROOF. Theorem 1 says that each of the integers

$$(2) \qquad\qquad 1, 2, \ldots, p - 1$$

has an order that is a divisor of $p - 1$. For each divisor d of $p - 1$, let $\psi(d)$ denote the number of integers in (2) that have order d. Restating what we have just said:

$$\sum_{d \mid p-1} \psi(d) = p - 1.$$

From Theorem 4 of Section 9, we have

$$(3) \qquad\qquad \sum_{d \mid p-1} \psi(d) = \sum_{d \mid p-1} \phi(d).$$

If we can show that $\psi(d) \leq \phi(d)$ for each d, it will follow from (3) that

$\psi(d) = \phi(d)$ for each d. In particular, the number of primitive roots of p will be $\psi(p - 1) = \phi(p - 1)$.

Choose some d. If $\psi(d) = 0$, then $\psi(d) < \phi(d)$ and we are done. If $\psi(d) \neq 0$, then there is an integer with order d; call it a. The congruence

(4) $$x^d \equiv 1 \pmod{p}$$

has, according to Lemma 4, exactly d solutions. Furthermore, (4) is satisfied by the d integers

(5) $$a, a^2, a^3, \ldots, a^d,$$

and because no two of these have the same least residue (mod p), they give all the solutions. From Lemma 2, the numbers in (5) that have order d are those powers a^k with $(k, d) = 1$. But there are $\phi(d)$ such numbers k. Hence $\psi(d) = \phi(d)$ in this case. As noted above, this completes the proof.

We have actually proved more than was stated in Theorem 4. Although we will not use what we have proved, it is worth stating as a

COROLLARY. If p is a prime and $d \mid (p - 1)$, then the number of least residues (mod p) with order d is $\phi(d)$.

Exercise 7. Use the table of powers (mod 7) at the beginning of this section to verify that the corollary is true for $p = 7$.

Theorem 4 does not actually help us to find a primitive root of a prime. To find one, we may use tables or trial. Here is a table giving the smallest positive primitive root, g_p, for each prime p less than 100.

p	2	3	5	7	11	13	17	19	23	29	31	37	41	43	47
g_p	1	2	2	3	2	2	3	2	5	2	3	2	6	3	5
p	53	59	61	67	71	73	79	83	89	97					
g_p	2	2	2	2	7	5	3	2	3	5					

No method is known for predicting what will be the smallest positive primitive root of a given prime p, nor is there much known about the distribution of the $\phi(p - 1)$ primitive roots among the least residues modulo p. For example, the primitive roots of 71 and 73 are

7	11	13	21	22	28		5	11	13	14	15	20
31	33	35	42	44	47		26	28	29	31	33	34
52	53	55	56	59	61		39	40	42	44	45	47
62	63	65	67	68	69		53	58	59	60	62	68

Primitive roots of 71	Primitive roots of 73

There are other numbers besides primes that have primitive roots. It can be proved that the only positive integers with primitive roots are 1, 2, 4, p^e, and $2p^e$, where p is an odd prime and e is a positive integer.

Exercise 8. Which of the integers 2, 3, . . . , 25 do not have primitive roots?

As an example of the application of primitive roots, we will use them to prove part of Wilson's Theorem quickly and elegantly. Let g be a primitive root of the odd prime p. From Theorem 3, we know that the least residues (mod p) of g, g^2, . . . , g^{p-1} are a permutation of $1, 2, . . . , p - 1$. Multiplying and using the fact that

$$1 + 2 + 3 + \cdots + (p - 1) = (p - 1)p/2$$

(see Appendix A, (p. 191) for a proof), we get

$$1 \cdot 2 \cdots (p - 1) \equiv g \cdot g^2 \cdots g^{p-1} \pmod{p}$$

or

$$(p - 1)! \equiv (g^p)^{(p-1)/2} \equiv g^{(p-1)/2} \pmod{p}.$$

But $g^{(p-1)/2}$ satisfies $x^2 \equiv 1 \pmod{p}$, and we know, either from Lemma 2 of Section 6 or from Lemma 4 of this section, that

$$g^{(p-1)/2} \equiv 1 \qquad \text{or} \qquad -1 \pmod{p}.$$

But the first case is impossible, since g is a primitive root of p. Thus $(p - 1)! \equiv -1 \pmod{p}$.

Logarithms are undeniably useful, and an analogous notion can be defined for integers modulo m, if m is an integer that has primitive roots. Let g be a primitive root of m. For k with $(k, m) = 1$, define an *index* of k (mod m), with respect to g, to be an integer t such that

(6) $$g^t \equiv k \pmod{m}.$$

Because m has primitive roots, such a t exists. By the *least index* of k (mod m), with respect to g (written ind$_g$ k), we will mean that integer t that satisfies (6) and is in addition a least residue (mod $\phi(m)$). Such a number exists because from (6) it follows that if t_1 and t_2 are indices of k (mod m), then $t_1 \equiv t_2$ (mod $\phi(m)$). For example, consider the primitive root 2 of 5. We have

$$2^0 \equiv 1, \qquad 2^1 \equiv 2, \qquad 2^2 \equiv 4, \qquad \text{and} \qquad 2^3 \equiv 3 \pmod{5},$$

so

$$\text{ind}_2 1 = 0, \qquad \text{ind}_2 2 = 1, \qquad \text{ind}_2 4 = 2, \qquad \text{and} \qquad \text{ind}_2 3 = 3.$$

Exercise 9. Calculate the least indices of the integers (mod 7), with respect to the primitive root 5.

Indices are useful in computation in the same way as logarithms:

THEOREM 5. *If m is a number with primitive roots and* ind$_g$ k *denotes the least index of* k *(mod m), with respect to the primitive root g, then*

(7) $\text{ind}_g \, ab \equiv \text{ind}_g \, a + \text{ind}_g \, b \ (mod \ \phi(m))$

(8) $\text{ind}_g \, a^n \equiv n \, \text{ind}_g \, a \ (mod \ \phi(m)).$

PROOF. Let $\text{ind}_g \, a = r$ and $\text{ind}_g \, b = s$. Then $g^r \equiv a$ and $g^s \equiv b$ (mod m). Thus

$$ab \equiv g^{r+s} \ (\text{mod } m).$$

Hence $r + s$ is an index of ab (mod m), and so

$$\text{ind}_g \, ab \equiv r + s \ (\text{mod } \phi(m)).$$

This proves statement (a). Statement (b) may be proved by repeated application of (a).

Exercise 10. Verify that, (mod 7),

$$\text{ind}_5 \, 2 + \text{ind}_5 \, 6 \equiv \text{ind}_5 \, 12,$$
$$6 \, \text{ind}_5 \, 2 \equiv \text{ind}_5 \, 64.$$

As with logarithms, indices turn multiplication problems into addition problems. If we have a table of least indices (mod m), with respect to any primitive root of m, then Theorem 5 allows us to solve congruences with comparative ease, just as tables of logarithms, to any base, simplify computations with real numbers. For example, using the primitive root 2 of 19, we can construct the following table of least indices:

n	1	2	3	4	5	6	7	8	9	10	11	12
$\text{ind}_2 \, n$	0	1	13	2	16	14	6	3	8	17	12	15
n	13	14	15	16	17	18						
$\text{ind}_2 \, n$	5	7	11	4	10	9						

We can use such a table to solve congruences. For example, if

$$13x \equiv 16 \ (\text{mod } 19),$$

then, from Theorem 5,

$$\text{ind}_2 \, 13x \equiv \text{ind}_2 \, 13 + \text{ind}_2 \, x \equiv \text{ind}_2 \, 16 \ (\text{mod } 18);$$

using the table,

$$\text{ind}_2 \, x \equiv 4 - 5 \equiv 17 \ (\text{mod } 18).$$

From the table again, we see that $x \equiv 10$ (mod 19). Another example, which would be difficult to solve with other methods, is

$$x^{13} \equiv 16 \ (\text{mod } 19).$$

From Theorem 5,

$$13 \text{ ind}_2 x \equiv \text{ind}_2 16 \equiv 4 \text{ (mod 18)}.$$

This has the solution $\text{ind}_2 x \equiv 10 \text{ (mod 18)}$, so from the table, $x \equiv 17 \text{ (mod 19)}$ satisfies the original congruence.

Problems

1. Which of the following integers have primitive roots?
   (a) 198, 199, 200, 201, 202, 203.
   (b) 10198, 10199, 10200, 10201, 10202, 10203.

2. What are the orders of 2, 4, 7, 8, 11, 13, and 14, modulo 15?

3. (a) Construct a table of least indices modulo 29 with respect to the primitive root 2.
   (b) Solve, using the table, $9x \equiv 2 \text{ (mod 29)}$.
   (c) Solve, using the table, $x^9 \equiv 2 \text{ (mod 29)}$.

4. Given that $\text{ind}_5 45 \equiv 45 \text{ (mod 96)}$, how many solutions are there to (a) $x^7 \equiv 45 \text{ (mod 97)}$, (b) $x^8 \equiv 45 \text{ (mod 97)}$, (c) $x^9 \equiv 45 \text{ (mod 97)}$?

5. (a) Show that 2 is a primitive root of 19.
   (b) Show that 2 is not a primitive root of 23.
   (c) Is 10 a primitive root of 11?

6. If $(a, m) \neq 1$, does there exist a positive integer t such that $a^t \equiv 1 \text{ (mod } m)$?

7. A says, "Look. These five pages of computation show that

$$457^{911} \equiv 1 \text{ (mod 10021)}."$$

   B says (after a glance at Table C, or perhaps Table A), "You err." Is B right?

8. If g is a primitive root of m, prove that $g^a \equiv g^b \text{ (mod } m)$ if and only if $a \equiv b \text{ (mod } \phi(m))$.

9. (a) Show that if g is a primitive root of p, then so is the least residue of g^k if $(k, p - 1) = 1$.
   (b) Find the twelve primitive roots of 37.

10. (a) Find all of the primitive roots of 11.
    (b) What is the index of 7 with respect to each?

11. If g is a primitive root of a prime p, show that $\text{ind}_g (-1) = (p - 1)/2$.

12. Suppose that 2 is a primitive root of an odd prime p and $\text{ind}_2 p - 1 = r$.
    (a) Show that $\text{ind}_2 p - 2 = r + 1$.
    (b) What is x if $\text{ind}_2 x = r + 2$?

13. Which integers have order 6, modulo 31?

14. If a has order e modulo p, show that

$$a^{e-1} + a^{e-2} + \cdots + 1 \equiv 0 \text{ (mod } p).$$

15. (a) If g and h are primitive roots of an odd prime p, then $g \equiv h^a \pmod{p}$ for some integer a. Show that a is odd.
    (b) Show that if g and h are primitive roots of an odd prime p, then the least residue of gh is not a primitive root of p.

16. Show that if a has order 3 $\pmod{p}$, then $a + 1$ has order 6 $\pmod{p}$.

17. Show that $131{,}071 = 2^{17} - 1$ is a prime.

18. Prove that if p and q are odd primes and $q \mid a^p + 1$, then either $q \mid a + 1$ or $q = 2kp + 1$ for some integer k.

19. Show that $(2^{19} + 1)/3$ is a prime.

20. If g is a primitive root of a prime p, show that it is impossible for $\text{ind}_g (n - 1)$, $\text{ind}_g n$, and $\text{ind}_g (n + 1)$ to be in arithmetic progression for any n.

21. (a) Show that if m is a number having primitive roots, then the product of the positive integers less than or equal to m and relatively prime to it is congruent to $-1 \pmod{m}$.
    (b) Show that the result in (a) is not always true if m does not have primitive roots.

22. The relation $(\log_a b)(\log_b a) = 1$ is well known.
    (a) What is the analogous statement for indices?
    (b) Prove it.

23. If $\log_a b = \log_b a$, then it is known that $b = a$ or $1/a$. Suppose that g and h are primitive roots of an odd prime p and $\text{ind}_g h = \text{ind}_h g$. What can be concluded?

Quadratic Congruences

After studying linear congruences, it is natural to look at quadratic congruences:

$$Ax^2 + Bx + C \equiv 0 \ (\text{mod } m).$$

In this section we will restrict the moduli to odd primes; in Appendix C we will see how to handle general quadratic congruences.

We will assume that $A \not\equiv 0 \ (\text{mod } p)$, because if $A \equiv 0 \ (\text{mod } p)$, then

(1) $$Ax^2 + Bx + C \equiv 0 \ (\text{mod } p)$$

would be a linear congruence, not a quadratic congruence. We know that there is an integer A' such that $AA' \equiv 1 \ (\text{mod } p)$. Hence (1) has the same solutions as

(2) $$x^2 + A'Bx + A'C \equiv 0 \ (\text{mod } p).$$

Exercise 1. Convert $2x^2 + 3x + 1 \equiv 0 \ (\text{mod } 5)$ to a quadratic whose first coefficient is 1.

If $A'B$ is even, we can complete the square in (2) to get

$$\left(x + \frac{A'B}{2}\right)^2 \equiv \left(\frac{A'B}{2}\right)^2 - A'C \ (\text{mod } p);$$

if $A'B$ is odd, we can change it to $p + A'B$, which is even, and then complete the square. In either case, we have replaced (1) with an equivalent congruence of the form

(3) $$y^2 \equiv a \ (\text{mod } p);$$

thus, if we can solve this congruence, we can solve any quadratic congruence (mod p).

Exercise 2. Change the quadratic in Exercise 1 to the form (3).

Exercise 3 (optional). By inspection, find all the solutions of the congruence in Exercise 2.

Such congruences do not always have solutions. For example, modulo 5,

$$0^2 \equiv 0, \qquad 1^2 \equiv 4^2 \equiv 1, \qquad \text{and} \qquad 2^2 \equiv 3^2 \equiv 4,$$

so $x^2 \equiv a \pmod 5$ has a solution for $a = 0$, 1, or 4 and no solution for $a = 2$ or 3. We note that $x^2 \equiv 0 \pmod p$ has only the solution $x \equiv 0 \pmod p$. We now show that if $p \nmid a$, then solutions of $x^2 \equiv a \pmod p$ come in pairs. This should be no surprise; since $r^2 = (-r)^2$, we have $r^2 \equiv (-r)^2 \pmod p$, so if r is a solution of $x^2 \equiv a \pmod p$, then so is the least residue $\pmod p$ of $-r$. Thus if r is a solution, so is $p - r$.

THEOREM 1. *Suppose that p is an odd prime. If $p \nmid a$, then $x^2 \equiv a \pmod p$ has exactly two solutions or no solutions.*

PROOF. Suppose that the congruence has a solution, and call it r. Then $p - r$ is a solution too, and it is different from r. (For if $r \equiv p - r \pmod p$, then $2r \equiv 0 \pmod p$; since $(2, p) = 1$, we get $r \equiv 0 \pmod p$, which is impossible.) Let s be any solution. Then $r^2 \equiv s^2 \pmod p$, whence $p \mid (r - s)(r + s)$. Thus

$$p \mid (r - s) \qquad \text{or} \qquad p \mid (r + s).$$

In the first case, $s \equiv r \pmod p$. In the second case, $s \equiv p - r \pmod p$. Since s, r, and $p - r$ are all least residues, we have $s = r$ or $p - r$; these are thus the only solutions.

This theorem is not true if the modulus is not prime. For example, $x^2 \equiv 1 \pmod 8$ has four solutions.

Exercise 4. If $p > 3$, what are the two solutions of $x^2 \equiv 4 \pmod p$?

It follows from Theorem 1 that if a is selected from the integers $1, 2, \ldots,$ $p - 1$, then $x^2 \equiv a \pmod p$ will have two solutions for $(p - 1)/2$ values of a and no solutions for the other $(p - 1)/2$ values of a. For example, $x^2 \equiv a \pmod 7$ has two solutions when $a = 1$, 2, or 4 and no solutions when $a = 3$, 5, or 6, as can be seen from the following table:

x	1	2	3	4	5	6
$x^2 \pmod 7$	1	4	2	2	4	1

Exercise 5. For what values of a does $x^2 \equiv a \pmod{11}$ have two solutions?

It would be nice to be able to tell the two groups apart. In this section we will do this by deriving *Euler's criterion:*

THEOREM 2. *If p is an odd prime and $p \nmid a$, then $x^2 \equiv a \pmod p$ has a solution or no solution according as*

$$a^{(p-1)/2} \equiv 1 \quad or \quad -1 \ (mod\ p).$$

First we introduce some new words: if $x^2 \equiv a \pmod m$ has a solution, then *a* is called a *quadratic residue* (mod *m*). If the congruence has no solutions, then *a* is called a *quadratic nonresidue* (mod *m*). There are also cubic residues, fourth-power residues and so on, but assuming there is no danger of confusion, we will omit the adjective quadratic and refer to residues and nonresidues for short.

Euler's criterion can be easily derived using primitive roots.

PROOF OF THEOREM 2. Let *g* be a primitive root of the odd prime *p*. Then $a \equiv g^k \pmod p$ for some *k*. If *k* is even, then $x^2 \equiv a \pmod p$ has a solution, namely the least residue of $g^{k/2}$; further, by Fermat's Theorem,

$$a^{(p-1)/2} \equiv (g^k)^{(p-1)/2} \equiv (g^{k/2})^{(p-1)} \equiv 1 \pmod p.$$

If *k* is odd, then

$$a^{(p-1)/2} \equiv (g^{(p-1)/2})^k \equiv (-1)^k \equiv -1 \pmod p,$$

and also $x^2 \equiv a \pmod p$ has no solution: if it did have one, say *r*, we would have

$$1 \equiv r^{p-1} \equiv (r^2)^{(p-1)/2} \equiv a^{(p-1)/2} \equiv -1 \pmod p,$$

which is impossible.

As an example of the application of the criterion, let us see if $x^2 \equiv 7 \pmod{31}$ has a solution. We must calculate $7^{(31-1)/2} = 7^{15}$ and see what its remainder is upon division by 31. Of course we do not need to carry out the actual division: we have

$$7^2 \equiv 49 \equiv 18 \pmod{31};$$

squaring, we get

$$7^4 \equiv 18^2 \equiv 324 \equiv 14 \pmod{31},$$
$$7^8 \equiv 14^2 \equiv 196 \equiv 10 \pmod{31},$$

and

$$7^{16} \equiv 10^2 \equiv 100 \equiv 7 \pmod{31}.$$

Since 7 and 31 are relatively prime, we may divide the last congruence by 7 to get $7^{15} \equiv 1 \pmod{31}$. It follows from Euler's criterion that $x^2 \equiv 7 \pmod{31}$ has a solution.

Though Euler's criterion tells us when $x^2 \equiv a \pmod p$ has solutions, it gives us no way of actually finding them. Of course, it is possible to substitute $x = 1, 2, 3, 4, \ldots$ until a solution is found, but this procedure can be long and tiresome. The following method—adding multiples of the modulus and factoring squares—is sometimes more convenient. For example, take $x^2 \equiv 7$ (mod 31), which we know to have a solution. Adding 31 repeatedly, we have

$$x^2 \equiv 7 \equiv 38 \equiv 69 \equiv 100 \equiv 10^2 \pmod{31},$$

and we see immediately that the congruence is satisfied when $x = 10$ or -10; the two solutions are thus 10 and 21. That example was easy; a more typical one is $x^2 \equiv 41 \pmod{61}$, which Euler's criterion shows to have a solution. We have

$$x^2 \equiv 41 \equiv 102 \equiv 163 \equiv 224 \equiv 4^2 \cdot 14 \pmod{61},$$

so

$$(x/4)^2 \equiv 14 \equiv 75 \equiv 5^2 \cdot 3 \pmod{61}.$$

Consequently,

$$(x/4 \cdot 5)^2 \equiv 3 \equiv 64 \equiv 8^2 \pmod{61},$$

and

$$x^2 \equiv (4 \cdot 5 \cdot 8)^2 \equiv 160^2 \equiv 38^2 \pmod{61}.$$

Thus $x \equiv \pm 38 \pmod{61}$, and the two solutions are 38 and 23. This method will always produce the solutions, with more or less labor.

Exercise 6. Find the solutions of $x^2 \equiv 8 \pmod{31}$.

Euler's criterion is sometimes cumbersome to apply, even to congruences with small numbers like $x^2 \equiv 3201 \pmod{8191}$. We will now develop a method for deciding when an integer is a quadratic residue (mod p). The method is relatively easy to apply, even when the numbers are 3201 and 8191. It is based on the famous quadratic reciprocity theorem, which has many applications other than the one we will use it for.

We start by introducing notation to abbreviate the long phrase, "$x^2 \equiv a$ (mod p) has a solution." We define the *Legendre symbol*, (a/p), where p is an odd prime and $p \nmid a$, in the following manner:

$(a/p) = 1$ if a is a quadratic residue (mod p),

$(a/p) = -1$ if a is a quadratic nonresidue (mod p).

For example, $(3/5) = -1$ because $x^2 \equiv 3 \pmod 5$ has no solutions, and

$(1/5) = 1$, because 1 is a quadratic residue (mod 5). Neither $(7/15)$ nor $(91/7)$ is defined, the first because the second entry in the symbol is not an odd prime, and the second because $7 \mid 91$.

Exercise 7. What is $(1/3)$? $(1/7)$? $(1/11)$? In general, what is $(1/p)$?

Exercise 8. What is $(4/5)$? $(4/7)$? $(4/p)$ for any odd prime p?

Exercise 9 (optional). Induce a theorem from the two preceding exercises.

To find out whether $x^2 \equiv 3201 \pmod{8191}$ has a solution, we can evaluate $(3201/8191)$. To do this, we will need some rules on how Legendre symbols can be manipulated. We will start with three simple but important properties.

THEOREM 3. *The Legendre symbol has the properties*

(A) *if $a \equiv b \pmod{p}$, then $(a/p) = (b/p)$,*

(B) *if $p \nmid a$, then $(a^2/p) = 1$,*

(C) *if $p \nmid a$ and $p \nmid b$, then $(ab/p) = (a/p)(b/p)$.*

In the above properties and throughout the rest of this section, we will agree that p and q represent odd primes, and that the first entry in a Legendre symbol is not a multiple of the second entry; with these conventions, all Legendre symbols are defined.

PROOF OF THEOREM 3. (A): Suppose that $x^2 \equiv a \pmod{p}$ has a solution. If $a \equiv b \pmod{p}$, then $x^2 \equiv b \pmod{p}$ also has a solution—the same one. This shows that

(4)   if $(a/p) = 1$ and $a \equiv b \pmod{p}$,   then $(b/p) = 1$.

Exercise 10. Verify that

(5)   if $(a/p) = -1$ and $a \equiv b \pmod{p}$, then $(b/p) = -1$.

Together, (4) and (5) show that (A) is true.

(B): Clearly, $x^2 \equiv a^2 \pmod{p}$ has a solution—namely, the least residue of $a \pmod{p}$.

(C): This important property of the Legendre symbol, in combination with the quadratic reciprocity theorem, makes the symbol useful for computations. In words, (C) says that the product of two residues is a residue; the product of two nonresidues is a residue; and the product of a residue and a nonresidue is a nonresidue. To prove (C) we use Euler's criterion: in terms of the Legendre symbol, it says

$$(a/p) = 1 \quad \text{if} \quad a^{(p-1)/2} \equiv 1 \pmod{p},$$

and

$$(a/p) = -1 \quad \text{if} \quad a^{(p-1)/2} \equiv -1 \pmod{p}.$$

Comparing the 1's and -1's, we see that

(6) $$(a/p) \equiv a^{(p-1)/2} \pmod{p}.$$

So, from (6) and the fact that $(xy)^n \equiv x^n y^n \pmod{p}$, we have

$$(ab/p) \equiv (ab)^{(p-1)/2} \equiv a^{(p-1)/2} b^{(p-1)/2} \equiv (a/p)(b/p) \pmod{p}.$$

We have not yet proved (C); we have only shown that

$$(ab/p) \equiv (a/p)(b/p) \pmod{p}.$$

But the left-hand side of the congruence is either 1 or -1, and so is the right-hand side. Hence, the only way that the two numbers can be congruent modulo p is if they are equal. We have now proved (C).

We can also use (6) to give quick proofs of (A) and (B). For example, to prove (B), we have from (6)

$$(a^2/p) \equiv (a^2)^{(p-1)/2} \equiv a^{p-1} \equiv 1 \pmod{p}$$

by Fermat's Theorem. Since $(a^2/p) = 1$ or -1, it follows that $(a^2/p) = 1$.

Exercise 11. Prove (A) with the help of (6).

Exercise 12. Prove that $(4a/p) = (a/p)$ for all a and p.

Exercise 13. Evaluate $(19/5)$ and $(-9/13)$ by using (A) and (B).

The quadratic reciprocity theorem tells us how (p/q) and (q/p) are related. The theorem was guessed by Euler years before it was first proved by Gauss, who eventually gave several proofs. This is an example of a deep and important theorem whose statement was arrived at by observation. Consider the following tables:

	5	7	11	13	17	19	23
3	-1	1	-1	1	-1	1	-1
5		-1	1	-1	-1	1	-1
7			1	-1	-1	-1	1
q 11				-1	-1	-1	-1
13					1	-1	1
17						1	-1
19							1

p

(p/q)

	5	7	11	13	17	19	23
3	-1	-1	1	1	-1	-1	1
5		-1	1	-1	-1	1	-1
7			-1	-1	-1	1	-1
q 11				-1	-1	1	1
13					1	-1	1
17						1	-1
19							-1

p

(q/p)

Can you by observation see any relation between (p/q) and (q/p)? These tables are perhaps too small to allow any firm guesses to be made, but note that the columns in both tables are the same for $p = 5$, 13, and 17. So are the rows in both tables the same for these three primes. What 5, 13, and 17 have that the rest of the primes less than 29 do not is the property of being congruent to 1 (mod 4). On this evidence, we might make the correct guess:

if either p or q is congruent to 1 (mod 4), then
$(p/q) = (q/p)$.

All of the entries not covered by this rule change sign from one table to the next. This behavior can be explained by the following hypothesis:

if p and q are both congruent to 3 (mod 4), then
$(p/q) = -(q/p)$.

These guesses are in fact generally true, and they make up

THEOREM 4 (*The quadratic reciprocity theorem*). *If p and q are odd primes and $p \equiv q \equiv 3$ (mod 4), then $(p/q) = -(q/p)$. Otherwise, $(p/q) = (q/p)$.*

We will postpone the proof of this theorem until the next section, but we will not hesitate to apply it for lack of a proof. Suppose that we want to see if $x^2 \equiv 85$ (mod 97) has a solution. That is, we want to evaluate $(85/97)$. With Theorems 3 and 4, we can carry the evaluation to a conclusion. We have

(7) $$(85/97) = (17 \cdot 5/97) = (17/97)(5/97)$$

by property (C) in Theorem 3. We will attack each factor in (7) separately. Because $97 \equiv 1$ (mod 4) (and, for that matter, $17 \equiv 1$ (mod 4) too), the quadratic reciprocity theorem says that

$$(17/97) = (97/17).$$

Property (A) in Theorem 3 says that

$$(97/17) = (12/17)$$

and

$$
\begin{aligned}
(12/17) &= (4 \cdot 3/17) = (4/17)(3/17) && \text{(by (C))} \\
&= (3/17) && \text{(by (B))} \\
&= (17/3) && \text{(by Theorem 4)} \\
&= (2/3) && \text{(by (A))} \\
&= -1 && \text{(by inspection).}
\end{aligned}
$$

The other factor is simpler:

$$(5/97) = (97/5) \qquad \text{(by Theorem 4)}$$
$$= (2/5) \qquad \text{(by (A))}$$
$$= -1 \qquad \text{(by inspection).}$$

Putting these calculations back in (7), we get

$$(85/97) = (17/97)(5/97) = (-1)(-1) = 1;$$

thus the congruence has a solution. By applying (A) first, we could have evaluated $(85/97)$ in another way:

$$(85/97) = (-12/97) = (-1/97)(4/97)(3/97) = (-1/97)(3/97).$$

We see that

$$(3/97) = (97/3) = (1/3) = 1,$$

so $(85/97) = (-1/97)$; if we know $(-1/97)$, we then know $(85/97)$.

If you look at a number of examples of Legendre symbols, it will become evident that to evaluate any one by using Theorems 3 and 4, it is enough to know what $(-1/p)$ and $(2/p)$ are for any p. Euler's criterion quickly gives us $(-1/p)$:

THEOREM 5. *If p is an odd prime, then*

$$(-1/p) = 1 \qquad if \qquad p \equiv 1 \ (mod \ 4),$$

and

$$(-1/p) = -1 \qquad if \qquad p \equiv 3 \ (mod \ 4).$$

In words, -1 is a quadratic residue of primes congruent to 1 (mod 4), and a nonresidue of all other odd primes.

PROOF. Euler's criterion says that

$$(-1/p) \equiv (-1)^{(p-1)/2} \ (mod \ p);$$

since $(p - 1)/2$ is even or odd according as p is congruent to 1 or 3 (mod 4), the theorem is proved.

In the example we were just considering, $(-1/97) = 1$ because $97 \equiv 1$ (mod 4).

Theorem 5 tells us that we can sometimes find square roots of -1 modulo p: whenever $p \equiv 1$ (mod 4), -1 has a square root (mod p).

Exercise 14. For which of the primes 3, 5, 7, 11, 13, 17, 19, and 23 is -1 a quadratic residue?

Exercise 15. Evaluate $(6/7)$ and $(2/23)(11/23)$.

It is not so easy to determine whether 2 has a square root (mod p). Euler's criterion says that

$$(2/p) \equiv 2^{(p-1)/2} \pmod{p},$$

but it is not obvious for which primes $2^{(p-1)/2}$ is congruent to 1 (mod p). We will find out in the next section. For now, we state the result:

THEOREM 6. *If p is an odd prime, then*

$$(2/p) = 1 \qquad if \qquad p \equiv 1 \quad or \quad 7 \ (mod\ 8),$$
$$(2/p) = -1 \qquad if \qquad p \equiv 3 \quad or \quad 5 \ (mod\ 8).$$

Theorem 6 together with Theorems 3 to 5 enable us to evaluate any Legendre symbol. For example, we can now evaluate $(3201/8191)$. The calculations go as follows:

$$(3201/8191) = (3/8191)(11/8191)(97/8191);$$
$$(3/8191) = -(8191/3) = -(1/3) = -1,$$
$$(11/8191) = -(8191/11) = -(7/11) = (11/7) = (4/7) = 1,$$

and

$$(97/8191) = (8191/97) = (43/97) = (97/43) = (11/43)$$
$$= -(43/11) = -(-1/11) = 1.$$

Thus we see that $(3201/8191) = (-1)(1)(1) = -1$. Compare the labor of evaluating $(3201/8191)$ as we have just done with that of determining by trial and error whether $x^2 \equiv 3201 \pmod{8191}$ has a solution. To calculate $1^2, 2^2, \ldots,$ 4095^2 and divide each by 8191 is no light task. Theorems 3 to 6 are, for this task at least, an enormous help.

Problems

1. Which of the following congruences have solutions?

(a) $x^2 \equiv 7 \pmod{53}$,   (c) $x^2 \equiv 14 \pmod{31}$,
(b) $x^2 \equiv 53 \pmod{7}$,   (d) $x^2 \equiv 625 \pmod{9973}$.

2. Find solutions for the congruences of Problem 1 that have them.

3. Solve

(a) $x^2 + x + 1 \equiv 0 \pmod{5}$,
(b) $x^2 + x \equiv 0 \pmod{5}$,
(c) $x^2 + x - 1 \equiv 0 \pmod{5}$.

4. (a) How many solutions has $x^2 \equiv 1 \pmod{16}$?

   (b) Does this contradict Theorem 1?

5. Solve

   (a) $2x^2 + 3x + 1 \equiv 0 \pmod 7$,

   (b) $3x^2 + x + 4 \equiv 0 \pmod 7$.

6. Which integers a, $1 \le a \le 30$, are quadratic residues $\pmod{31}$?

7. Calculate, modulo 23,

      (a) 2^{11},   (b) 3^{11},   (c) 4^{11},   (d) 5^{11},   (e) 22^{11},   (f) 21^{11}.

8. For which of $p = 3, 5, 7, 11, 13$, and 17 is $x^2 \equiv -2 \pmod p$ solvable?

9. Calculate

   (a) $(33/71)$, (b) $(34/71)$, (c) $(35/71)$, (d) $(36/71)$.

10. Calculate (a) $(1234/4567)$, (b) $(4321/4567)$.

11. Show that if $p = q + 4a$ (p and q odd primes), then $(p/q) = (a/q)$.

12. Show that if $p = 12k + 1$ for some k, then $(3/p) = 1$.

13. Does $x^2 \equiv 53 \pmod{97}$ have a solution? Does $x^2 \equiv 97 \pmod{53}$?

14. Show that Theorem 6 could also be written

$$(2/p) = (-1)^{(p^2 - 1)/8}.$$

15. Prove, using Legendre symbols, that the product of three quadratic non-residues $\pmod p$ is a nonresidue $\pmod p$. What is the product of two non-residues and a residue?

16. The quadratic reciprocity theorem is sometimes stated as follows: If p and q are odd primes, then

$$(p/q)(q/p) = (-1)^{(p-1)(q-1)/4}.$$

Is it the same theorem?

17. A says, "I bet 7 divides $n^2 + 1$ for some n." B says, "You're on." Who wins?

18. Show that if a is a quadratic residue $\pmod p$ and $ab \equiv 1 \pmod p$, then b is a quadratic residue $\pmod p$.

19. Does $x^2 \equiv 211 \pmod{159}$ have a solution?

20. Generalize Problem 18 by finding what condition on r will guarantee that if a is a quadratic residue $\pmod p$ and $ab \equiv r \pmod p$, then b is a quadratic residue $\pmod p$?

21. Suppose that $p = q + 4a$, where p and q are odd primes. Show that $(a/p) = (a/q)$.

Quadratic Reciprocity

In this section we will prove the two theorems stated and used without proof in the last section: the quadratic reciprocity theorem, and the theorem that enables us to evaluate $(2/p)$. The proofs are not easy, especially the proof of the quadratic reciprocity theorem. At the base of both theorems is the following result, sometimes called *Gauss's Lemma:*

THEOREM 1. *Suppose that p is an odd prime, $p \nmid a$, and there are among the least residues (mod p) of*

$$a, 2a, 3a, \ldots, \left(\frac{p-1}{2}\right) a$$

exactly g that are greater than $(p-1)/2$. Then $x^2 \equiv a \pmod{p}$ has a solution or no solution according as g is even or odd. Otherwise stated,

$$(a/p) = (-1)^g.$$

Before proving the theorem, we will illustrate its application by taking $a = 5$ and $p = 17$. We have $(p-1)/2 = 8$, and the integers

$$5, 10, 15, 20, 25, 30, 35, 40$$

have least residues (mod 17)

$$5, 10, 15, 3, 8, 13, 1, 6.$$

Three of these are greater than $(p-1)/2$. Theorem 1 then says that 5 is a quadratic nonresidue (mod 17), which is so.

Exercise 1. Check that the theorem gives the right result in this case by applying Euler's criterion and showing that $5^8 \equiv -1 \pmod{17}$.

PROOF OF THEOREM 1. Let

$$r_1, r_2, \ldots, r_k$$

denote those among the least residues (mod p) of

$$a, 2a, \ldots, ((p-1)/2)a$$

that are less than or equal to $(p-1)/2$, and let

$$s_1, s_2, \ldots, s_g$$

denote those that are greater than $(p-1)/2$. Thus $k + g = (p-1)/2$. To prove the theorem, it is enough, by Euler's criterion, to show that

$$a^{(p-1)/2} \equiv (-1)^g \pmod{p},$$

and this is what we proceed to do. (In the example above, $k = 5$, and $g = 3$; the set of r's is $\{5, 3, 8, 1, 6\}$, and the set of s's is $\{10, 15, 13\}$.) Both in the example and in general, no two of the r's are congruent (mod p). Suppose that two were. Then we would have for some k_1 and k_2,

$$k_1 a \equiv k_2 a \pmod{p}, \qquad 0 \le k_1 \le (p-1)/2, \quad 0 \le k_2 \le (p-1)/2.$$

Because $(a, p) = 1$, it follows that $k_1 = k_2$. For the same reason, no two of the s's are congruent (mod p). Now, consider the set of numbers

(1) $$r_1, r_2, \ldots, r_k, p - s_1, p - s_2, \ldots, p - s_g.$$

Each integer n in the set satisfies $1 \le n \le (p-1)/2$, and there are $(p-1)/2$ elements in the set. Gauss noticed, and we will now prove, that the numbers in the set are all different. From this it will follow that the elements in (1) are just a permutation of the integers

(2) $$1, 2, \ldots, (p-1)/2,$$

and thus the product of the elements in (1) is the same as the product of the elements in (2). From this the theorem will follow. In the example we considered, the set of r's was $\{5, 3, 8, 1, 6\}$, and the set of $(p-s)$'s was $\{7, 2, 4\}$; between them, they include all the integers from 1 to 8.

To show that the elements in (1) are distinct, we have only to show that

$$r_i \not\equiv p - s_j \pmod{p}$$

for any i and j, because we have already seen that the r's and s's are distinct among themselves. Suppose that for some i and j we have

$$r_i \equiv p - s_j \pmod{p}.$$

Then $r_i + s_j \equiv 0 \pmod{p}$. Since $r_i \equiv ta \pmod{p}$ and $s_j \equiv ua \pmod{p}$ for some t and u with t and u positive integers less than or equal to $(p-1)/2$, we would have

$$(t + u)a \equiv 0 \pmod{p};$$

since $(a, p) = 1$, we have $t + u \equiv 0 \pmod{p}$, and this is impossible, because $2 \leq t + u \leq p - 1$. Thus all of the elements in the set (1) are distinct, and consequently are a rearrangement of the elements in (2). Hence,

$$(3) \qquad r_1 r_2 \cdots r_k (p - s_1)(p - s_2) \cdots (p - s_g) = 1 \cdot 2 \cdots ((p - 1)/2).$$

Because $p - s_j \equiv -s_j \pmod{p}$ for all j, and because there are g such terms, (3) becomes

$$(4) \qquad r_1 r_2 \cdots r_k s_1 s_2 \cdots s_g (-1)^g \equiv \left(\frac{p-1}{2}\right)! \pmod{p}.$$

But $r_1, r_2, \ldots, r_k, s_1, s_2, \ldots, s_g$ are, by definition, the least residues $\pmod{p}$ of

$$a, 2a, \ldots, ((p-1)/2)a$$

in some order, so that the product $r_1 r_2 \cdots r_k s_1 s_2 \cdots s_g$ is congruent $\pmod{p}$ to $a(2a)(3a) \cdots ((p-1)/2)a$. Thus (4) gives

$$a^{(p-1)/2}(-1)^g \left(\frac{p-1}{2}\right)! \equiv \left(\frac{p-1}{2}\right)! \pmod{p}.$$

The common factor is relatively prime to p and may be cancelled to give

$$a^{(p-1)/2}(-1)^g \equiv 1 \pmod{p}.$$

If we multiply both sides of the last congruence by $(-1)^g$, we have

$$a^{(p-1)/2} \equiv (-1)^g \pmod{p}.$$

But we know that $a^{(p-1)/2} \equiv (a/p) \pmod{p}$. Putting the last two congruences together, and noting that if the two numbers are congruent $\pmod{p}$, then they must be equal, we have

$$(a/p) = (-1)^g,$$

and this is what we wanted to prove.

Exercise 2. Apply the theorem to determine whether $x^2 \equiv 7 \pmod{23}$ has a solution.

We will now apply the last theorem to evaluate $(2/p)$ for any odd prime p. According to the theorem, we need to find out how many of the least residues $\pmod{p}$ of

(5)
$$2, 4, 6, \ldots, 2\left(\frac{p-1}{2}\right)$$

are greater than $(p-1)/2$. Since all the numbers in (5) are already least residues, none of them being larger than p, we have only to see how many of them are greater than $(p-1)/2$. Let the first even integer greater than $(p-1)/2$ be $2a$. Then

(6)
$$\frac{p-1}{2} < 2a \leq \frac{p-1}{2} + 2.$$

Exercise 3. Note that the number g we are looking for is
$$g = \frac{p-1}{2} - (a-1).$$

Exercise 4. Use (6) to show that g is determined by
$$\frac{p-1}{4} \leq g < \frac{p-1}{4} + 2.$$

Exercise 5. Verify that the entries in the following table are correct.

p	3	5	7	11	13	17	19	23	29
g	1	1	2	3	3	4	5	6	7
$(-1)^g$	-1	-1	1	-1	-1	1	-1	1	-1

After verifying the entries, you may have noticed for which primes g is even and for which it is odd. If not, suppose that $p \equiv 1 \pmod 8$. Then $p = 1 + 8k$ for some k, and $(p-1)/4 = 2k$. Hence $g = 2k$, and g is even. So, if $p \equiv 1 \pmod 8$, then $(2/p) = 1$. Similarly, if $p = 3 + 8k$ for some k, then $(p-1)/4 = 2k + 1/2$, g is odd, and $(2/p) = -1$.

Exercise 6. Check the cases $p = 8k + 5$ and $p = 8k + 7$.

Exercise 7. We do not need to consider $p = 8k$, $8k + 2$, $8k + 4$, or $8k + 6$. Why not?

Thus we have proved

THEOREM 2. *If p is an odd prime, then*

$$(2/p) = 1 \qquad if \qquad p \equiv 1 \quad or \quad 7 \pmod 8,$$
$$(2/p) = -1 \qquad if \qquad p \equiv 3 \quad or \quad 5 \pmod 8.$$

As an example of the use of Theorem 2, we will state and prove a result that is a digression, but one that is pleasing and perhaps a little surprising. Although we know when a number has primitive roots, finding the actual roots is, in general, not easy. For example, 2 is a primitive root of 3, 5, 11, 13, 19, 29, 37, 53, 59, 61, 67, and 83 among the primes less than 100, and it is not a primitive root of the others. No theorem has been proved that will predict the primes of which 2 is a primitive root, and it has not even been proved that 2 is a primitive root of infinitely many primes. But we do have

THEOREM 3. *If p and 4p + 1 are both primes, then 2 is a primitive root of 4p + 1.*

PROOF. Let $q = 4p + 1$. Then $\phi(q) = 4p$, so 2 has order 1, 2, 4, p, $2p$, or $4p$ (mod q); we want to show that the first five cases are impossible. We have

$$2^{2p} \equiv 2^{(q-1)/2} \equiv (2/q) \pmod{q}$$

by Euler's criterion. But p is odd, so $4p \equiv 4 \pmod 8$, and $q \equiv 4p + 1 \equiv 5$ (mod 8); we know from Theorem 2 that 2 is a quadratic nonresidue of primes congruent to 5 (mod 8). Hence

$$2^{2p} \equiv -1 \pmod{q},$$

so 2 does not have order $2p$. Nor can the order of 2 be any of the divisors of $2p$, which are of course 1, 2, and p. Since 2 does not have order 4 either ($2^4 \equiv 1 \pmod{q}$ implies $q \mid 15$, so $q = 5$, which is impossible), the theorem is proved.

We will now give Gauss's third proof of the quadratic reciprocity theorem. Those with a developed sense of mathematical esthetics find this proof almost breathtakingly elegant. It depends on Gauss's Lemma (Theorem 1) and on the lemma we will now prove.

LEMMA 1. *If p and q are different odd primes, then*

$$\sum_{k=1}^{(p-1)/2} \left[\frac{kq}{p} \right] + \sum_{k=1}^{(q-1)/2} \left[\frac{kp}{q} \right] = \frac{p-1}{2} \cdot \frac{q-1}{2}.$$

The notation $[kq/p]$ denotes the greatest integer not larger than kq/p. If this notation is unfamiliar, see Appendix B (p. 196).

Exercise 8. Verify that the lemma is true for $p = 5$ and $q = 7$.

PROOF OF THE LEMMA. The idea of the proof is geometrical. It is an example of how analytic geometry can dramatically simplify some proofs. For convenience in notation, let

$$S(p, q) = \sum_{k=1}^{(p-1)/2} \left[\frac{kq}{p} \right];$$

we are thus trying to prove that

$$S(p, q) + S(q, p) = (p - 1)(q - 1)/4.$$

As noted in Appendix B, $[kq/p]$ is the number of integers in the interval $0 < x \le kq/p$. Hence $[kq/p]$ is equal to the number of *lattice points* (that is, points whose coordinates are integers) that lie on the line $x = k$ and below the line $y = qx/p$ and on or above the line $y = 1$. Looking at the figure for $k = 1, 2, \ldots, (p - 1)/2$, we see that

$$S(p, q) = \sum_{k=1}^{(p-1)/2} [kq/p]$$

is the number of lattice points inside or on the boundary of the polygon $ABCD$.

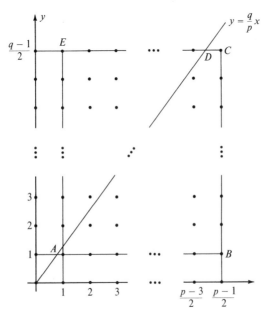

In the same way, we can see that $S(q, p)$ is the number of lattice points inside or on the boundary of the polygon ADE. Since $(q, p) = 1$, there are no lattice points on the segment AD. It follows that $S(p, q) + S(q, p)$ is the number of lattice points inside or on the boundary of the whole rectangle with vertices at $(1, 1)$, B, C, and E. This number is easy to count: it is $((p - 1)/2)\cdot((q - 1)/2)$, and this proves the lemma.

THEOREM 4 (the quadratic reciprocity theorem). *If p and q are odd primes, then*

$$(p/q)(q/p) = (-1)^{(p-1)(q-1)/4}.$$

Note that this is equivalent to the way we stated the theorem earlier (Theorem 4 of Section 11): If $p \equiv q \equiv 3 \pmod 4$, then $(p/q) = -(q/p)$; otherwise, $(p/q) = (q/p)$. This is so because $(p - 1)(q - 1)/4$ is even unless $p \equiv q \equiv 3 \pmod 4$.

PROOF. As in the proof of Gauss's Lemma, let us take the least residues (mod p) of

$$q, 2q, 3q, \ldots, \frac{p-1}{2}q$$

and separate them into two classes. Put those less than or equal to $(p-1)/2$ in one class and call them

$$r_1, r_2, \ldots, r_k,$$

and put those greater than $(p-1)/2$ in another and call them

$$s_1, s_2, \ldots, s_g.$$

Thus $k + g = (p-1)/2$. The conclusion of Gauss's Lemma was that $(q/p) = (-1)^g$. For ease in notation, let

$$R = r_1 + r_2 + \cdots + r_k \qquad \text{and} \qquad S = s_1 + s_2 + \cdots + s_g.$$

In the course of proving Gauss's Lemma, we showed that the set of numbers

(7) $r_1, r_2, \ldots, r_k, p - s_1, p - s_2, \ldots, p - s_g$

was a permutation of

(8) $1, 2, \ldots, (p-1)/2.$

It follows that the sum of the elements in (7) is the same as the sum of the elements in (8). Remember that in the proof of Gauss's Lemma, we took the *product* of the elements in (7) and equated it to the product of the elements in (8). Here, then, is a possible starting point of this proof: Gauss may have thought (in German), "What would happen if I equated the sums instead of the products?" and then constructed the proof. Whatever it was that he thought, the lesson to be learned is that proofs often do not start at the beginning. The sum of the numbers in (8) is, by a well-known formula $(1 + 2 + \cdots + n = n(n+1)/2$; see Appendix A, p. 191, for a proof),

$$\frac{1}{2}\left(\frac{p-1}{2}\right)\left(\frac{p-1}{2} + 1\right) = \frac{p^2 - 1}{8}.$$

The sum of the elements in (7) is

$$\sum_{j=1}^{k} r_j + \sum_{j=1}^{g} (p - s_j) = R + gp - S.$$

Thus we have

(9) $R = S - gp + (p^2 - 1)/8.$

The least residue (mod p) of jq ($j = 1, 2, \ldots, (p - 1)/2$) is the remainder when we divide jq by p. We know the quotient, $[jq/p]$, so if we let t_j denote the least residue (mod p) of jq, we have

$$jq = [jq/p]p + t_j,$$

$j = 1, 2, \ldots, (p - 1)/2$. If we sum these equations over j, we have

$$\sum_{j=1}^{(p-1)/2} jq = \sum_{j=1}^{(p-1)/2} [jq/p]p + \sum_{j=1}^{(p-1)/2} t_j,$$

or

$$q \sum_{j=1}^{(p-1)/2} j = p \sum_{j=1}^{(p-1)/2} [jq/p] + \sum_{j=1}^{k} r_j + \sum_{j=1}^{g} s_j,$$

or

(10) $$q(p^2 - 1)/8 = pS(p, q) + R + S.$$

Substituting into this from (9), we get

$$q(p^2 - 1)/8 = pS(p, q) + 2S - gp + (p^2 - 1)/8,$$

or

(11) $$(q - 1)(p^2 - 1)/8 = p(S(p, q) - g) + 2S.$$

In (11), the left-hand side is even (because $(p^2 - 1)/8$ is an integer and $q - 1$ is even), and $2S$ is even. It follows that the remaining term in (11) is even, and so $S(p, q) - g$ is even. Hence

$$(-1)^{S(p,q)-g} = 1.$$

Since $(-1)^g = (q/p)$, then

(12) $$(-1)^{S(p,q)} = (-1)^g = (q/p).$$

Now we can repeat the argument with p and q interchanged—nowhere have we required q to have a property that p does not—and get

(13) $$(-1)^{S(q,p)} = (p/q).$$

Multiplying (12) and (13), we have

$$(-1)^{S(p,q)+S(q,p)} = (p/q)(q/p),$$

and from Lemma 1, we have

$$(-1)^{(p-1)(q-1)/4} = (p/q)(q/p),$$

which is what we wanted.

Problems

1. Adapt the method used in the text to evaluate $(2/p)$ to evaluate $(3/p)$.

2. Show that 3 is a quadratic nonresidue of all primes of the form $4^n + 1$.

3. Show that 3 is a quadratic nonresidue of all Mersenne primes greater than 3.

4. (a) Prove that if $p \equiv 7$ (mod 8), then $p \mid (2^{(p-1)/2} - 1)$.
   (b) Find a factor of $2^{83} - 1$.

5. (a) If p and $q = 10p + 3$ are odd primes, show that $(p/q) = (3/p)$.
   (b) If p and $q = 10p + 1$ are odd primes, show that $(p/q) = (-1/p)$.

6. (a) Which primes can divide $n^2 + 1$ for some n?
   (b) Which odd primes can divide $n^2 + n$ for some n?
   (c) Which odd primes can divide $n^2 + 2n + 2$ for some n?

7. (a) Show that if $p \equiv 3$ (mod 4) and a is a quadratic residue (mod p), then $p - a$ is a quadratic nonresidue (mod p).
   (b) What if $p \equiv 1$ (mod 4)?

8. If $p > 3$, show that p divides the sum of its quadratic residues.

9. (a) Suppose that $p \geq 5$ is prime. Show that -3 is a quadratic residue (mod p) if $p \equiv 1$ or 7 (mod 12) and a nonresidue if $p \equiv 5$ or 11 (mod 12).
   (b) Suppose that p is an odd prime, $p \neq 3$, and $p \nmid a$. Suppose that $x^3 \equiv a$ (mod p) has a solution r. Then

   $$(x - r)(x^2 + xr + r^2) \equiv 0 \text{ (mod } p\text{)}.$$

   Show that $x^2 + xr + r^2 \equiv 0$ (mod p) has two solutions different from r if and only if $p \equiv 1$ or 7 (mod 12).
   (c) If $p \geq 5$, show that the number of distinct nonzero cubic residues (mod p) is

   $$\begin{array}{llll} p - 1 & \text{if} & p \equiv 5 & \text{or} \quad 11 \text{ (mod 12)} \\ (p - 1)/3 & \text{if} & p \equiv 1 & \text{or} \quad 7 \text{ (mod 12)}. \end{array}$$

10. If p is an odd prime, evaluate

   $$(1 \cdot 2/p) + (2 \cdot 3/p) + \cdots + ((p - 2)(p - 1)/p).$$

11. Show that if $p \equiv 1$ (mod 4), then $x^2 \equiv 1$ (mod p) has a solution given by the least residue (mod p) of $((p - 1)/2)!$.

Numbers in Other Bases

One of the great accomplishments of the human mind, and one that made mathematics possible, was the invention of our familiar notation for writing integers. We write integers in a place-value notation, with each place indicating a different power of 10. For example,

$$314,159 = 3 \cdot 10^5 + 1 \cdot 10^4 + 4 \cdot 10^3 + 1 \cdot 10^2 + 5 \cdot 10^1 + 9 \cdot 10^0.$$

There is no reason why some integer other than 10 could not be used for the same purpose. The choice of 10 is only an anatomical accident. In fact, other integers—we will call them *bases*—have been used in the past. The Babylonians of 3,000 years ago sometimes used the base 60, and the ancient Mayans used the base 20. Today, numbers written in the bases 2, 8, and 16 are used by computers. In this section, we will look at integers in bases other than 10.

We start by looking at a special case.

THEOREM 1. *Every positive integer can be written as a sum of distinct powers of two.*

For example, $22 = 2^4 + 2^2 + 2^1$ and $23 = 2^4 + 2^2 + 2^1 + 2^0$. But $24 = 2^3 + 2^3 + 2^3$ is not a proper representation, because the powers of two are not distinct.

Exercise 1. Write 31 and 33 as sums of distinct powers of two.

PROOF OF THEOREM 1. The idea of the proof is to take an integer n and subtract from it the largest power of 2 that is smaller than it—say 2^k. Then we do the same for $n - 2^k$. If we continue this process, we will eventually get a representation of n in the form that we want. To make this argument rigorous, we prove the theorem by induction: $1 = 2^0$, $2 = 2^1$, and $3 = 2^0 + 2^1$, so the theorem is true if the integer is 1, 2, or 3. Suppose, now, that every integer k, $k \leq n - 1$, can be written as a sum of distinct powers of 2. We want to show that n can also be so written. We know that n falls between some two distinct powers of 2; that is, there is an integer r such that

(1) $2^r \leq n < 2^{r+1}$.

Exercise 2. What is r if $n = 74$? If $n = 174$?

The largest power of 2 not larger than n is 2^r. Let $n' = n - 2^r$. Then $n' \leq n - 1$, so the induction assumption tells us that it can be written as the sum of distinct powers of two:

$$n' = 2^{e_1} + 2^{e_2} + \cdots + 2^{e_k},$$

where $e_i \neq e_j$ if $i \neq j$. Since $n' = n - 2^r$, we have

(2) $n = 2^r + 2^{e_1} + 2^{e_2} + \cdots + 2^{e_k}$,

and so n can be written as a sum of powers of two. To complete the proof, we need to show that r is different from any of $e_1, e_2, \ldots, e_k$.

Exercise 3. Why is this so?

We now show that the representation (2) is unique.

THEOREM 2. *Every positive integer can be written as the sum of distinct powers of 2 in at most one way.*

PROOF. Suppose that n has two representations as a sum of distinct powers of 2. We will show that the representations are really the same. To make the notation less clumsy, we note that any sum of distinct powers of 2 can be written in the form

(3) $d_0 + d_1 \cdot 2 + d_2 \cdot 2^2 + d_3 \cdot 2^3 + \cdots + d_k \cdot 2^k$

for some k, where $d_i = 0$ or 1 for each i. Conversely, every such sum is a sum of distinct powers of 2. Hence it is immaterial whether we write n in the form (2) or (3), and (3) has the advantage of having no subscripts on the exponents. If n has two representations, we have

(4)
$$\begin{aligned} n &= d_0 + d_1 \cdot 2 + d_2 \cdot 2^2 + \cdots + d_k \cdot 2^k \\ &= e_0 + e_1 \cdot 2 + e_2 \cdot 2^2 + \cdots + e_k \cdot 2^k, \end{aligned}$$

where $d_i = 0$ or 1 and $e_i = 0$ or 1 for each i. (Note that we lose no generality in assuming that the two representations have the same number of terms. If one is longer than the other, we can add zero terms to the shorter one until the two have the same length.) Subtracting the second representation in (4) from the first gives

(5) $\quad 0 = (d_0 - e_0) + (d_1 - e_1) \cdot 2 + (d_2 - e_2) \cdot 2^2 + \cdots + (d_k - e_k) \cdot 2^k.$

Hence $2 \mid (d_0 - e_0)$. But since d_0 and e_0 are either 0 or 1, it follows that

$$-1 \le d_0 - e_0 \le 1.$$

Since the only multiple of 2 in that range is zero, $d_0 = e_0$. Thus the first term in (5) disappears, and we may divide what remains by 2 to get

(6) $\quad 0 = (d_1 - e_1) + (d_2 - e_2) \cdot 2 + \cdots + (d_k - e_k) \cdot 2^{k-1}.$

The same argument as before shows that $d_1 = e_1$. Dropping the term $d_1 - e_1$ from (6), dividing by 2 and applying the same argument again, we get $d_2 = e_2$. And so on: $d_3 - e_3 = d_4 - e_4 = \cdots = d_k - e_k = 0$, and the two representations in (4) were the same.

Theorems 1 and 2 show that every n can be written in exactly one way in the form

(7) $$d_0 + d_1 \cdot 2 + d_2 \cdot 2^2 + d_3 \cdot 2^3 + \cdots + d_k \cdot 2^k$$

for some k, where each d is either 0 or 1. This is like the ordinary decimal representation for integers. It is so like it that we can write numbers of the form (7) in the same style as we usually write integers. The powers of 2 and the plus signs in (7) are not essential, since the thing that determines n is the sequence $d_0, d_1, \ldots, d_k$. We will write the expression (7) as

(8) $$(d_k d_{k-1} \cdots d_1 d_0)_2,$$

and say that we have written the integer *in the base* 2. The subscript 2 reminds us that d_r is multiplied by 2^r. For example,

$$101001_2 = 1 + 0 \cdot 2 + 0 \cdot 2^2 + 1 \cdot 2^3 + 0 \cdot 2^4 + 1 \cdot 2^5$$
$$= 1 + 8 + 32 = 41.$$

In the other direction,

$$94 = 64 + 16 + 8 + 4 + 2$$
$$= 1 \cdot 2^6 + 0 \cdot 2^5 + 1 \cdot 2^4 + 1 \cdot 2^3 + 1 \cdot 2^2 + 1 \cdot 2^1 + 0 \cdot 2^0$$
$$= 1011110_2.$$

Exercise 4. Evaluate 1001_2, 111_2, and 1000000_2.

Exercise 5. Write 2, 20, and 200 in the base 2.

We know that every integer can be uniquely expressed in the form

$$d_0 + d_1 \cdot 10 + d_2 \cdot 10^2 + \cdots + d_k \cdot 10^k$$

for some k, with $0 \leq d_i < 10$, $i = 0, 1, \ldots, k$, and Theorems 1 and 2 show that every integer can be uniquely expressed in the form

$$d_0 + d_1 \cdot 2 + d_2 \cdot 2^2 + \cdots + d_k \cdot 2^k$$

for some k, with $0 \leq d_i < 2$, $i = 0, 1, \ldots, k$. What we can do for 2 and 10, we ought to be able to do for any integer greater than 1. In fact, we can. We will prove

THEOREM 3. *Let $b \geq 2$ be any integer (called the base). Any positive integer can be written uniquely in the base b; that is, in the form*

(9) $$n = d_0 + d_1 \cdot b + d_2 \cdot b^2 + \cdots + d_k \cdot b^k$$

for some k, with $0 \leq d_i < b$, $i = 0, 1, \ldots, k$.

PROOF. We will first show that each integer has such a representation and then show that it is unique. To show that there is a representation, we could adapt the proof of Theorem 1, but the proof that we will present here (which could also be applied to Theorem 1) gives a construction for the digits of n in the base b. Divide n by b: the division algorithm says

$$n = q_1 b + d_0, \qquad 0 \leq d_0 < b.$$

We can divide the quotient by b,

$$q_1 = q_2 b + d_1, \qquad 0 \leq d_1 < b,$$

and continue the process,

$$q_2 = q_3 b + d_2, \qquad 0 \leq d_2 < b,$$
$$q_3 = q_4 b + d_3, \qquad 0 \leq d_3 < b,$$

and so on. Since $n > q_1 > q_2 > q_3 > \cdots$ and each q_i is nonnegative, the sequence of q's will sooner or later terminate. That is, we will come to k such that

$$q_k = 0 \cdot b + d_k \qquad 0 \leq d_k < b.$$

But then

$$n = d_0 + q_1 b = d_0 + (d_1 + q_2 b)b = d_0 + d_1 b + q_2 b^2$$
$$= d_0 + d_1 b + (d_2 + q_3 b)b^2 = d_0 + d_1 b + d_2 b^2 + q_3 b^3$$

$$\cdots$$

$$= d_0 + d_1 b + d_2 b^2 + \cdots + q_k b^k$$
$$= d_0 + d_1 b + d_2 b^2 + \cdots + d_k b^k,$$

and this is the desired representation.

To show that it is unique, we use the same idea we used in the proof of Theorem 2. Suppose that n has two representations:

$$n = d_0 + d_1b + d_2b^2 + \cdots + d_kb^k$$
$$= e_0 + e_1b + e_2b^2 + \cdots + e_kb^k$$

for some k, where

(10) $$0 \le d_i < b \quad \text{and} \quad 0 \le e_i < b$$

for $i = 0, 1, \ldots, k$. (As in Theorem 2, there is no loss in generality in assuming that the two representations have the same number of terms.) Subtracting one representation from the other gives

$$0 = (d_0 - e_0) + (d_1 - e_1)b + (d_2 - e_2)b^2 + \cdots + (d_k - e_k)b^k.$$

We see that $b \mid (d_0 - e_0)$. From (10) it follows that $d_0 = e_0$.

Exercise 6. Complete the proof.

For short, we will write

$$d_0 + d_1b + \cdots + d_kb^k = (d_kd_{k-1} \cdots d_1d_0)_b.$$

For example,

$$111_7 = 1 + 1 \cdot 7 + 1 \cdot 7^2 = 57_{10}.$$

(We will usually omit the subscript b when $b = 10$. Unless noted otherwise, every integer without a subscript is written in base 10.)

To find the representation of a base-10 integer in the base b, the scheme used in the proof of Theorem 3 is as good as any. For example, to write 31415 in the base 8, we perform repeated divisions by 8:

$$31415 = 8 \cdot 3926 + 7,$$
$$3926 = 8 \cdot 490 + 6,$$
$$490 = 8 \cdot 61 + 2,$$
$$61 = 8 \cdot 7 + 5,$$
$$7 = 8 \cdot 0 + 7,$$

and hence $31415_{10} = 75267_8$. (*Check:*

$$75267_8 = 7 + 6 \cdot 8 + 2 \cdot 8^2 + 5 \cdot 8^3 + 7 \cdot 8^4 = 7 + 48 + 128 + 2560 + 28762$$
$$= 31415.)$$

To make the arithmetic easier, the divisions may be arranged differently. For example, we have $31415_{10} = 160406_7$:

	quotients	remainder
7)	31415	
	4487	6
	641	0
	91	4
	13	0
	1	6
	0	1

Problems

1. Write 1492 in the base (a) 2, (b) 3, (c) 7, (d) 9, (e) 11.

2. Calculate
(a) 3141_5, (b) 3141_6, (c) 3141_7, (d) 3141_9.

3. Solve for x: (a) $123_4 = x_5$, (b) $234_5 = x_6$, (c) $123_x = 1002_4$.

4. Verify that the entries in the following addition table in the base 7 are correct, and complete it.

+	1	2	3	4	5	6	10
1	2	3	4	5	6	10	11
2	3	4	5	6	10	11	12
3	4	5	6	10	11	12	13
4	5	6	10	11	12	13	14
5							
6							
10							

5. Verify that the entries in the following multiplication table in the base 7 are correct, and complete it.

·	2	3	4	5	6	10
2	4	6	11	13	15	20
3	6	12	15	21	24	30
4	11	15	22	26	33	40
5						
6						

6. All numbers in this problem are in the base 7. Calculate

(a) $15 + 24 + 33$,
(b) $314 + 152 + 265 + 351$,
(c) $42 \cdot 12$,
(d) $314 \cdot 152$.

7. Evaluate as rational numbers in the base 10

$\qquad$ (a) $(.25)_7,$ $\qquad$ (b) $(.333\ldots)_7,$ $\qquad$ (c) $(.545454\ldots)_7.$

(By $(.d_1d_2d_3\ldots)_b$ we mean $d_1/b + d_2/b^2 + d_3/b^3 + \cdots.$)

8. In which bases b is 1111_b divisible by 5?

9. (a) Show that $123_7, 132_7, 312_7, 231_7, 321_7,$ and 213_7 are even integers.

$\quad$ (b) Show that in the base 7, an integer is even if and only if the sum of its digits is even.

$\quad$ (c) In which other bases is it true that if an integer is even, then any permutation of its digits is even?

10. (a) Show that $121_3 = 4^2,\ 121_4 = 5^2,$ and $121_5 = 6^2.$

$\quad$ (b) Guess and prove a theorem.

$\quad$ (c) Evaluate 169_b in base 10 $(b \geq 10).$

11. If a and b are positive integers, then a^2 ends with an even number of zeros, and $10b^2$ ends with an odd number of zeros. Hence $10b^2 = a^2$ is impossible in nonzero integers, and it follows that $10^{1/2}$ is irrational.

$\quad$ (a) Adapt the above proof to integers in the base b (b not a square) to show that $b^{1/2}$ is irrational.

$\quad$ (b) Show that $b^{1/m}$ (b not an mth power, $m = 3, 4, \ldots$) is irrational by the same argument.

12. Consider the following lists:

	List 1				List 2				List 4		
1	9	17	25	2	10	18	26	4	12	20	28
3	11	19	27	3	11	19	27	5	13	21	29
5	13	21	29	6	14	22	30	6	14	22	30
7	15	23	31	7	15	23	31	7	15	23	31

	List 8				List 16		
8	12	24	28	16	20	24	28
9	13	25	29	17	21	25	29
10	14	26	30	18	22	26	30
11	15	27	31	19	23	27	31

Pick a number from 1 to 31—any number—and see which of the above lists it appears in. If you add the numbers of the lists in which the number appears, you will get the number you picked. Why does this trick work?

13. Prove that every positive integer can be written uniquely in the form

$$n = e_0 + 3e_1 + 3^2e_2 + 3^3e_3 + \cdots + 3^ke_k$$

for some k, where $e_i = -1, 0,$ or $1, i = 0, 1, \ldots, k.$

14. An eccentric philanthropist undertakes to give away \$100,000. He is eccentric because he insists that each of his gifts be a number of dollars that is a power of two, and he will give no more than one gift of any amount. How does he distribute the money?

15. (a) Prove that every positive odd integer can be represented in the form

$$n = d_0 + d_1 \cdot 2 + d_2 \cdot 2^2 + \cdots + d_k \cdot 2^k,$$

where $d_i = 1$ or $-1, i = 0, 1, \ldots, k,$ but the representation is not unique.

(b) A and B play a series of games, the first for $1, the second for $2, and so on, the stakes being doubled each time. After eight games, A is $31 ahead. Which games did he win?

16. If weights can be placed in both pans of a balance, what is the minimum number of weights needed to weigh objects weighing 1, 2, . . . , 100 pounds?

SECTION **14**

Duodecimals

In this section, we will take a close look at arithmetic in the base 12. In doing this, we will re-examine the familiar processes of addition, subtraction, multiplication, and division in an unfamiliar setting. After completing the exercises, you should realize how expert you really are at computation (in the base 10), and what enormous labor it takes to become quick at arithmetical operations. We all worked hard in the third grade. Any base would serve as well as 12 to give practice, but some parts of arithmetic—notably decimals— are nicer in the base 12 than they are in the base 10. Besides, there is a good deal of twelveness in everyday life: items are measured by the dozen and gross, there are 12 months in the year, 12 inches in a foot, half a dozen feet in a fathom, two dozen hours in the day, and 30 dozen degrees in the circle. The reason for this abundance of twelves is the easy divisibility of 12 by 3, 4, and 6; we want to make such divisions much more often than we want to divide things by 5. Counting by tens is the result of a really unfortunate accident; how much better ordered the world would be if we had six fingers on each hand! Because we don't, it is unlikely that we will ever abandon counting by tens, even though counting by dozens is manifestly better. But there is a Duodecimal Society of America that devotes itself to educating the public in preparation for the day when the change is made. Although the public is still largely untouched, the Society looks forward to that day with faint but unquenchable hope.

In order to count by dozens, we need two new digits to represent 10_{10} and 11_{10}. From now on in this section, all numbers will be *duodecimals*— that is, written in the base 12, unless otherwise indicated with a subscript.

The settled notation among duodecimalists for 10_{10} and 11_{10} seems to be χ and ε, the Greek letters chi and epsilon. But they are pronounced "dec" and "el." Thus duodecimal counting goes

$$1, 2, 3, 4, 5, 6, 7, 8, 9, \chi, \varepsilon, 10, 11, \ldots, 1\chi, 1\varepsilon, 20, \ldots,$$
$$30, \ldots, 40, \ldots, \chi 0, \ldots, \varepsilon 0, \ldots, 100, \ldots.$$

We need names for duodecimal numbers. The Duodecimal Society advocates "do" (from dozen) for 10 and "gro" for 100, so that, for example, 15 is "do five" and 327 is "3 gro 2 do 7." Unfortunately, the Society's names for larger and smaller multiples of a dozen, of which a dozen examples are given here,

10	Do	.1	Edo
100	Gro	.01	Egro
1000	Mo	.001	Emo
10000	Do-mo	.0001	Edo-mo
100,000	Gro-mo	.00001	Egro-mo
1,000,000	Bi-mo	.000001	Ebi-mo

may sound just close enough to baby talk to give you the giggles (2, 201,110 is "2 bi-mo, 2 gro-mo, mo, gro, do"), but .37χ5 can always be read "point three seven dec five," and if you want to call 1χ and 5ε "decteen" and "fifty-el" instead of "do dec" and "five do el," most probably only the most fanatical duodecimalists would object.

Addition of duodecimals is not hard if we remember to carry one whenever we sum to a dozen. Here is the addition table for six:

6	6	6	6	6	6	6	6	6	6	6	6
1	2	3	4	5	6	7	8	9	χ	ε	10
7	8	9	χ	ε	10	11	12	13	14	15	16

And here are some summations:

5	31	123	$\chi\chi\chi$
4	41	456	$\varepsilon\,\varepsilon\,\varepsilon$
3	15	789	$\chi\,\varepsilon\chi$
2	9	$\chi\,\varepsilon 0$	$\varepsilon\chi\,\varepsilon$
12	94	2036	3996

Exercise 1. Calculate $9 + 4$, $\chi + \varepsilon$, $\varepsilon 1 + 1\varepsilon$, and $16 + 19 + 37$.

Although we can carry the addition tables in our heads without too much trouble, we need a multiplication table to look at when we multiply, just as when we learned to multiply in the base χ.

Duodecimal Multiplication Table

·	2	3	4	5	6	7	8	9	χ	ε	10
2	4	6	8	χ	10	12	14	16	18	1χ	20
3	6	9	10	13	16	19	20	23	26	29	30
4	8	10	14	18	20	24	28	30	34	38	40
5	χ	13	18	21	26	2ε	34	39	42	47	50
6	10	16	20	26	30	36	40	46	50	56	60
7	12	19	24	2ε	36	41	48	53	5χ	65	70
8	14	20	28	34	40	48	54	60	68	74	80
9	16	23	30	39	46	53	60	69	76	83	90
χ	18	26	34	42	50	5χ	68	76	84	92	χ0
ε	1χ	29	38	47	56	65	74	83	92	χ1	ε0

Exercise 2. Verify that the χ-times table is correct.

With the aid of the table, multiplication is no problem. For example,

$$
\begin{array}{r} 34 \\ 5 \\ \hline 148 \end{array}
\qquad
\begin{array}{r} 1755 \\ \chi \\ \hline 14262 \end{array}
\qquad
\begin{array}{r} \chi\chi \\ \varepsilon\varepsilon \\ \hline 9\,\varepsilon2 \\ 9\,\varepsilon2 \\ \hline \chi912 \end{array}
$$

Exercise 3. Calculate $14\cdot2$, $14\cdot3$, and $9\cdot\chi\cdot\varepsilon$.

With enough practice, we could absorb the entries of the table and learn to do without it; eventually, "9 times 9" would produce "6 do 9" purely by reflex.

Division is slightly harder, even with the aid of the table, because lack of experience may lead us to choose the wrong digit in a quotient. It takes practice to be able to see at a glance how many χ5's there are in 763.

Exercise 4. How many are there?

Here are some worked-out divisions:

$$
\begin{array}{r} 5\,)\,456\,(\,\chi8 \\ \underline{42} \\ 36 \\ \underline{34} \\ 2 \end{array}
\qquad
\begin{array}{r} 22\,)\,456\,(\,20 \\ \underline{44} \\ 16 \end{array}
\qquad
\begin{array}{r} 31\,)\,4159\,(\,140 \\ \underline{31} \\ 105 \\ \underline{104} \\ 19 \end{array}
$$

Exercise 5. Calculate 1966/6 and 1111/5.

In duodecimals, 1/3 and 1/6 have terminating expansions—$1/3 = 4/10 = .4$ and $1/6 = 2/10 = .2$—and this is more pleasant than the case in the base χ. The expansion of 1/5, however, does *not* terminate:

$$
\begin{array}{r}
.2497 \\
5\overline{)1.0000} \\
\underline{\chi} \\
20 \\
\underline{18} \\
40 \\
\underline{39} \\
30 \\
\underline{2\,\varepsilon} \\
1
\end{array}
$$

So $1/5 = .24972497\ldots$. We will signify the repeating part of a repeating decimal by putting a bar over it; thus we will write $1/5 = .\overline{2497}$. We will continue to call such things "decimals," even though some other name ("doimals," perhaps?) would be more appropriate.

Exercise 6. Calculate the decimal representation of $1/7$.

Here is a table of reciprocals up through one decth and one elth (or, in other words, through dec edo and el edo).

n	2	3	4	5	6	7	8
$1/n$	.6	.4	.3	$.\overline{2497}$	.2	$.\overline{186\chi35}$	.16

n	9	χ	ε
$1/n$	.14	$.\overline{12497}$	$.\overline{1}$

Half of these terminate, and one elth, like one ninth in the base χ, has a particularly simple repeating part.

To convert from decimals to rational numbers, we use the same principles as in base χ. For example, $.25 = 25/100$. Is this fraction in its lowest terms? Neither numerator nor denominator is divisible by five, and because of our inexperience with duodecimals, we may find it hard to recognize any common factor, or the lack of one. Here is a short factor table:

2	prime	11	prime	$21 = 5^2$	
3	prime	$12 = 2\cdot7$		$22 = 2\cdot11$	
$4 = 2^2$		$13 = 3\cdot5$		$23 = 3^3$	
5	prime	$14 = 2^4$		$24 = 2^2\cdot7$	
$6 = 2\cdot3$		15	prime	25	prime
7	prime	$16 = 2\cdot3^2$		$26 = 2\cdot3\cdot5$	
$8 = 2^3$		17	prime	27	prime
$9 = 3^2$		$18 = 2^2\cdot5$		$28 = 2^5$	
$\chi = 2\cdot5$		$19 = 3\cdot7$		$29 = 3\cdot\varepsilon$	
ε	prime	$1\chi = 2\cdot\varepsilon$		$2\chi = 2\cdot15$	
$10 = 2^2\cdot3$		1ε	prime	$2\varepsilon = 5\cdot7$	
		$20 = 2^3\cdot3$		$30 = 2^2\cdot3^2$	

Since 25 is a prime, it follows that 25/100 is in its lowest terms.

Repeating decimals can be converted to fractions in the usual way. For example, let $N = .6666\ldots$. Then $10N = 6.6666\ldots$, so $10N - N = 6$. Thus $\varepsilon N = 6$, and $N = 6/\varepsilon$; this is a fraction in its lowest terms.

Problems

1. Continue the factor table in the text to 40.

2. Calculate (a) $3141 + 5926$, (b) $3141 - 5926$, (c) $3141 \cdot 5926$, (d) $3141/5926$ to three places.

3. Calculate (a) 2^{10}, (b) $9!$, (c) $1/3^3$.

4. Show that $.292929\ldots = 3/11$.

5. Take your age, add 11, double the result, then multiply by 6 and subtract 110. Show that the number you calculate is 10 times your age.

6. Which is more, $\$4\varepsilon.\varepsilon6$ or $\$(59.95)_\chi$?

7. If x is 0, 1, 4, or 9, show that the last digit of x^n is x for $n = 2, 3, \ldots$.

8. Show that the last digit of a square is 0, 1, 4, or 9.

9. Show that the last digit of x^n, $n = 2, 3, \ldots$

   (a) is 0 if $x = 6$
   (b) is 4 if $x = \chi$
   (c) is x if $x = 3, 5, 7, 8$, or ε and n is odd.

χ. From Problems 7, 8, and 9, conclude that the last digit of x^n, $n = 2, 3, \ldots$ is never 2, 6, or χ for any x.

ε. With which digits can a prime end?

10. Show that any integer whose last digit is 3, 6, 9, or 0 has 3 for a factor.

11. Show that any integer whose last digit is 4, 8, or 0 has 4 for a factor, and that any integer whose last digit is 6 or 0 has 6 for a factor.

12. (a) True or false? $19^2 = 361$. $21^2 = 441$. $23^2 = 529$.
    (b) Can you explain the phenomenon in (a)?

13. All of 5, 15, 25, 35, and 45 are prime. Can this go on?

14. Write as rational numbers in lowest terms

$$\text{(a) } 3.14 \qquad \text{(b) } .090909\ldots$$

15. Write as repeating decimals (a) $22/7$, (b) $1/11$.

16. Find $2^{1/2}$ correct to two places.

17. Show that any integer whose digits sum to a multiple of ε is divisible by ε.

18. Using the fact that $1001 = 7 \cdot 11 \cdot 17$, develop tests for the divisibility of an integer by 7, 11, and 17.

19. The Duodecimal Society of America also advocates the do-metric system of weights and measures: 1000 yards to the mile, 10 ounces to the pound, and 10 flounces to the pint. The Society relates distance, weight, and volume by requiring that a cubic yard hold 1000 pints of water, which weigh 1000 pounds. If we keep the yard as it is now, how do the do-metric mile, pint, and pound compare to the ordinary mile, pint, and pound?

1χ. (a) How many days are there in the year?

(b) What other three-digit numbers have the same property—that is, $d_1d_2d_3 = ((d_1 + 1)d_2d_3)_\chi$?

(c) Are there any four-digit numbers with this property?

Decimals

Some fractions have nice decimal expansions ($1/8 = .125$); others are not as nice, but still tolerable ($1/3 = .333 \ldots$); and others are not nice at all ($1/17 = .0588235294117647 \ldots$). In this section we will see which fractions are nicest, and we will find a way of determining how long the repeating part of a repeating decimal is without having to carry out the actual calculation of the decimal. In doing so, we will use nothing deeper than the division algorithm and some congruences.

We will denote

$$d_1/10 + d_2/10^2 + d_3/10^3 + \cdots \qquad (0 \leq d_k < 10)$$

by

$$.d_1d_2d_3 \ldots.$$

A bar over part of a decimal will indicate that this part repeats indefinitely. For example,

$$.01\overline{47} = .0147474747 \ldots \qquad \text{and} \qquad 1/3 = .\overline{3}.$$

Exercise 1. Write $.01\overline{47}$ as a rational number.

Exercise 2. Write $7/41$ as a decimal with a bar over its repeating part.

Let us make a table of the decimal expansions of the reciprocals of the first few integers and see if we can notice any pattern in them. A zero in the period column means that the decimal terminates.

n	$1/n$	Period		n	$1/n$	Period
2	.5	0		16	.0625	0
3	$.\overline{3}$	1		17	$.\overline{0588235294117647}$	16
4	.25	0		18	$.0\overline{5}$	1
5	.2	0		19	$.\overline{052631578947368421}$	18
6	$.1\overline{6}$	1		20	.05	0
7	$.\overline{142857}$	6		21	$.\overline{047619}$	6
8	.125	0		22	$.0\overline{45}$	2
9	$.\overline{1}$	1		23	$.\overline{0434782608695652173913}$	22
10	.1	0		24	$.041\overline{6}$	1
11	$.\overline{09}$	2		25	.04	0
12	$.08\overline{3}$	1		26	$.0\overline{384615}$	6
13	$.\overline{076923}$	6		27	$.\overline{037}$	3
14	$.0\overline{714285}$	6		28	$.03\overline{571428}$	6
15	$.0\overline{6}$	1		29	$.\overline{0344827586206896551724137931}$	28

The integers in the table whose decimal reciprocals terminate are

$$2, 4, 5, 8, 10, 16, 20, 25,$$

and one thing these numbers have in common is that they are all of the form 2^a5^b for some nonnegative integers a and b. We might guess that the decimal expansion of the reciprocal of any number of this form terminates. The next three such numbers are 32, 40, and 50, and

$$1/32 = .03125, \quad 1/40 = .025, \quad 1/50 = .02$$

all terminate. In fact, this guess is right.

THEOREM 1. *If a and b are any nonnegative integers, then the decimal expansion of $1/2^a5^b$ terminates.*

PROOF. Let M be the maximum of a and b. Then

$$10^M(1/2^a5^b) = 2^{M-a}5^{M-b}$$

is an integer—call it n. Clearly, $n \leq 10^M$. Thus

$$\frac{1}{2^a5^b} = \frac{n}{10^M},$$

so the decimal expansion of $1/2^a5^b$ consists of the digits of n, perhaps preceded by some zeros.

Exercise 3. Calculate M for 16, 20, and 25, and compare with the table to see that it gives the correct length of the expansion.

Exercise 4. How many places are there in the expansions of 1/128, 1/320, and 1/800?

The converse of Theorem 1 is also true.

THEOREM 2. *If $1/n$ has a terminating decimal expansion, then $n = 2^a 5^b$ for some nonnegative integers a and b.*

PROOF. Let the terminating decimal expansion of $1/n$ be

$$1/n = .d_1 d_2 \cdots d_k$$
$$= d_1/10 + d_2/10^2 + \cdots + d_k/10^k.$$

Then

$$1/n = (d_1 10^{k-1} + d_2 10^{k-2} + \cdots + d_k)/10^k.$$

Call the integer in parentheses m. Then the last equation is

$$1/n = m/10^k \quad \text{or} \quad mn = 10^k.$$

The only prime divisors of 10^k are 2 and 5, and so the only prime divisors of n are 2 and 5. This proves the theorem.

Theorems 1 and 2 completely take care of terminating decimals. Among the expansions in the table that do not terminate are some with long periods, including $n = 17, 19, 23,$ and 29. In each case, n is a prime and the period of $1/n$ is $n - 1$. But not all primes p have period $p - 1$ for their reciprocals: $1/13$ has period 6, not 12, and $1/11$ has period 2, not 10. As a first step in investigating the lengths of the periods of reciprocals, we prove

THEOREM 3. *The length of the decimal period of $1/n$ is no longer than $n - 1$.*

PROOF. Let t be such that $10^t < n < 10^{t+1}$. Then using the division algorithm repeatedly, we have

(1)
$$
\begin{array}{ll}
10^{t+1} = d_1 n + r_1, & 0 < r_1 < n, \\
10 r_1 = d_2 n + r_2, & 0 \le r_2 < n, \\
10 r_2 = d_3 n + r_3, & 0 \le r_3 < n, \\
\cdots, & \cdots, \\
10 r_k = d_{k+1} n + r_{k+1}, & 0 \le r_{k+1} < n, \\
\cdots, & \cdots.
\end{array}
$$

Note that each d_k is less than 10, because for $k = 2, 3, \ldots,$

$$d_k n = 10 r_{k-1} - r_k \le 10 r_{k-1} < 10n,$$

and

$$d_1 n = 10^{t+1} - r_1 < 10^{t+1} = 10 \cdot 10^t < 10n.$$

From (1) we see that

(2)
$$r_k/n = d_{k+1}/10 + r_{k+1}/10n.$$

If we divide both sides of the first equation in (1) by $10^{t+1}n$ and apply (2) over and over, we get

$$
\begin{aligned}
1/n &= d_1/10^{t+1} + r_1/n10^{t+1} \\
&= d_1/10^{t+1} + d_2/10^{t+2} + r_2/n10^{t+2} \\
&= d_1/10^{t+1} + d_2/10^{t+2} + d_3/10^{t+3} + r_3/n10^{t+3} \\
&\qquad \cdots \\
&= d_1/10^{t+1} + d_2/10^{t+2} + d_3/10^{t+3} + d_4/10^{t+4} + \cdots ,
\end{aligned}
$$

and thus $d_1, d_2, d_3 \ldots$ are the digits in the decimal expansion of $1/n$. For example, for $n = 7$ we have

$$
\begin{aligned}
10 &= 1 \cdot 7 + 3, \\
30 &= 4 \cdot 7 + 2, \\
20 &= 2 \cdot 7 + 6, \\
60 &= 8 \cdot 7 + 4, \\
40 &= 5 \cdot 7 + 5, \\
50 &= 7 \cdot 7 + 1, \\
10 &= 1 \cdot 7 + 3, \\
&\cdots ,
\end{aligned}
$$

and so the decimal expansion of $1/7$ is $.\overline{142857}$.

Each of the remainders $r_1, r_2, \ldots$ may assume one of the n values $0, 1, 2, \ldots, n - 1$. Hence, among the $n + 1$ integers $r_1, r_2, \ldots, r_{n+1}$, there must be two that are equal. (If you put $n + 1$ objects in n boxes, one box will contain two objects.) If $r_j = r_k$, then it follows from (1) that $d_{k+1} = d_{j+1}$, $d_{k+2} = d_{j+2}, \ldots$, and the decimal repeats, with period no greater than n.

Exercise 5. Apply the division algorithm to find the decimal expansion of $1/41$.

If n is relatively prime to 10, we can get more information about the period of $1/n$.

THEOREM 4. *If $(n, 10) = 1$, then the period of $1/n$ is r, where r is the smallest positive integer such that $10^r \equiv 1 \ (mod \ n)$.*

PROOF. We first note that the integer r exists. The least residues (mod n) of $1, 10, 10^2, 10^3, \ldots, 10^{n-1}$ may assume only the values $1, 2, 3, \ldots, n - 1$, because no power of 10 is divisible by n. Again we have n objects to be put in $n - 1$ boxes: there exist nonnegative integers a and b, $a \neq b$ and both smaller than n, such that $10^a \equiv 10^b \ (mod \ n)$. Dividing the congruence by the smaller power of 10, which is possible since $(n, 10) = 1$, will give r.
Since $10^r \equiv 1 \ (mod \ n)$, we know that

(3) $$10^r - 1 = kn$$

for some integer k. Written in the base 10, k has at most r digits (because $k < 10^r$). Let

$$k = d_{r-1}d_{r-2} \cdots d_1d_0 = d_{r-1}10^{r-1} + d_{r-2}10^{r-2} + \cdots + d_110 + d_0,$$

where $0 \le d_k < 10$ for $k = 0, 1, \ldots, r$. Then from (3)

$$\frac{1}{n} = \frac{k}{10^r - 1} = \frac{d_{r-1}d_{r-2} \cdots d_0}{10^r} \cdot \frac{1}{1 - 10^{-r}}$$

$$= (.d_{r-1}d_{r-2} \cdots d_0)(1 + 10^{-r} + 10^{-2r} + \cdots)$$

$$= \overline{.d_{r-1}d_{r-2} \cdots d_0}.$$

This shows that the period of $1/n$ is *at most* r. (The sequence

$$123123123123. \ldots$$

repeats after every ninth digit, but its period is three.) We must show that the period is no smaller than r. Suppose that the period of $1/n$ is s, with $s < r$. We will show that $10^s \equiv 1 \pmod n$, and this contradicts the assumption that r is the smallest positive integer such that $10^r \equiv 1 \pmod n$. The division algorithm in (1) says that for any k,

$$10r_k \equiv r_{k+1} \pmod n.$$

Thus

$$10^2 r_k \equiv 10r_{k+1} \equiv r_{k+2} \pmod n,$$

and in general,

$$10^t r_k \equiv r_{k+t} \pmod n$$

for any positive t. If $1/n$ is periodic with period s, then $d_{k+s} = d_k$ for all sufficiently large k, and hence $r_{k+s} = r_k$: this is not obvious enough, and more detail is needed.

$$10^s r_k \equiv r_{k+s} \equiv r_k \pmod n$$

for all sufficiently large k. If we can show that $(r_k, n) = 1$ for all k, it will follow that $10^s \equiv 1 \pmod n$, and the theorem will be proved. But from

$$10r_{k-1} = d_k n + r_k$$

it follows that if $p \mid r_k$ and $p \mid n$, then $p \mid 10r_{k-1}$. Since $(10, n) = 1$, we have $p \mid r_{k-1}$ for all k. But from

$$10^t = d_1 n + r_1$$

we get $p \mid 10$, which is impossible. Hence $(r_k, n) = 1$, and the theorem is proved.

The foregoing discussion could have been phrased in the language of primitive roots, with a gain in elegance and a loss in length. But we have not done this, following instead the principle that a cannon should not be used to shoot flies.

Exercise 6. Apply Theorem 4 to find the period of $1/41$.

Unfortunately, there are no general rules for looking at an integer and discovering what the number r is. Even among the primes, those for which $r = p - 1$ (7, 17, 19, 23, 29, . . .) are scattered in a pattern no one has yet been able to decipher.

So far we have considered reciprocals only. But the general rational number is no more difficult. It is clear that if we multiply a fraction by an integer that cancels no factors in its denominator, we do not change the period of its decimal expansion. The same holds for division by any number of the form $2^a 5^b$. Hence $1/2^a 5^b n$ has the same period as $1/n$, and if $(c, n) = 1$, then c/n has the same period as $1/n$. With these remarks, Theorem 4 lets us determine the period of the decimal expansion of c/n for any n not of the form $2^a 5^b$:

THEOREM 5. *If* $n \neq 2^a 5^b$ *and* $(c, n) = 1$, *then the period of the decimal expansion of* c/n *is* r, *the smallest positive integer such that*

$$10^r \equiv 1 \ (mod \ n_1),$$

where

$$n = 2^a 5^b n_1$$

and $(n_1, 10) = 1$.

From Theorem 5 and Lemma 1 of Section 10 it follows that the period of c/n must be a divisor of $\phi(n_1)$.

Problems

1. Find the periods of the expansions of
   (a) $1/66$, (b) $1/666$, (c) $1/4608$, (d) $1/925$, (e) $1/101$, (f) $1/1001$.

2. Find the smallest positive integer r such that $10^r \equiv 1 \pmod{n}$ if n is (a) 33, (b) 37.

3. A says, "With enormous labor, I have calculated the length of the period of $1/31415$. It is exactly 15707." B says, "You made a mistake." Is B correct?

4. Fill in any missing details in the argument in the text that $1/n$ and $1/2^b 5^b n$ have the same period for any nonnegative integers a and b.

5. Prove that if the decimal expansion of $1/n$ in the base b (that is,
$$1/n = d_1/b + d_2/b^2 + d_3/b^3 + \cdots, \qquad 0 \le d_k < b)$$
terminates, then every prime divisor of n is a divisor of b.

6. Prove that if every prime divisor of n is a divisor of b, then the decimal expansion of $1/n$ in the base b terminates.

7. Which of the reciprocals of 13, 14, . . . , 25 have terminating decimal expansions in the base 12?

8. Prove that if $(n, b) = 1$, then the period of the decimal expansion of $1/n$ in the base b is the smallest positive integer such that $b^r \equiv 1 \pmod{n}$.

9. In the base 2, what is the period of the decimal expansion of (a) 1/3, (b) 1/5, (c) 1/7, (d) 1/9, (e) 1/11?

10. Calculate the decimal expansions in the base 2 of (a) 1/3, (b) 1/5, (c) 1/9.

11. In the base 12, what is the period of the decimal expansion of (a) 1/7, (b) 1/11, (c) 1/17?

12. Calculate the decimal expansion in the base 12 of (a) 1/13, (b) 1/14.

13. Calculate the following decimal expansions.

(a) $1/9^2$ in the base 10.
(b) $1/7^2$ in the base 8.
(c) $1/6^2$ in the base 7.
(d) Guess a theorem.
(e) Prove it.

14. Show that

$$\sum_{n=1}^{\infty} 7^{-n(n+1)/2}$$

is irrational.

Pythagorean Triangles

More than 3500 years ago the Babylonians knew that the triangle whose sides have length 120, 119, and 169 is a right triangle. They knew many other such triangles too, including those with sides

$$4800, 4601, 6649,$$
$$360, \ 319, \ 481,$$
$$6480, 4961, 8161,$$
$$2400, 1679, 2929,$$
$$2700, 1771, 3229,$$

but it is not clear to what use they were put. In this section we will determine all such triangles.

The famous *Pythagorean Theorem*, which was first proved around 2500 years ago, states that if x and y are the legs of a right triangle and z its hypotenuse, then

$$x^2 + y^2 = z^2.$$

Many proofs are known, including one constructed by James Garfield (before he was President of the United States). Mathematical knowledge is thus not a disqualification for high office. A right triangle whose sides are integers we will call a *Pythagorean triangle*. (Strictly speaking, the sides are not integers; they are line segments whose lengths are denoted by integers, but no confusion can arise that is not willful.) The problem of finding all Pythagorean triangles is the same as that of finding all solutions in positive integers of

(1) $$x^2 + y^2 = z^2.$$

We first note that we may as well assume that x and y are relatively prime. Suppose not: let $x^2 + y^2 = z^2$ and $(x, y) = d$. Then $d \mid z$, and

$$(x/d)^2 + (y/d)^2 = (z/d)^2,$$

and we also know that $(x/d, y/d) = 1$. This shows that any solution of (1) may be derived from a solution in which the terms on the left are relatively prime, by multiplication by a suitable factor. Thus when we find all solutions of $x^2 + y^2 = z^2$ with $(x, y) = 1$, we will be able to find all solutions of $x^2 + y^2 = z^2$.

Exercise 1. If $(x, y) = 1$ and $x^2 + y^2 = z^2$, show that $(y, z) = (x, z) = 1$.

We will call a solution $x = a$, $y = b$, $z = c$ of $x^2 + y^2 = z^2$ in which a, b, and c are positive and $(a, b) = 1$ a *fundamental solution*. From Exercise 1 it follows that if a, b, c is a fundamental solution, then no two of a, b, c, have a common factor.

LEMMA 1. *If a, b, c is a fundamental solution of $x^2 + y^2 = z^2$, then exactly one of a and b is even.*

PROOF. The integers a and b cannot both be even in a fundamental solution.

Exercise 2. Why not?

Nor can a and b both be odd. Suppose that they were. Then $a^2 \equiv 1 \pmod 4$ and $b^2 \equiv 1 \pmod 4$. Thus

$$c^2 = a^2 + b^2 \equiv 2 \pmod 4,$$

which is impossible. The only possibility left is that one of a and b is even and the other is odd.

COROLLARY. *If a, b, c is a fundamental solution, then c is odd.*

PROOF. $a^2 + b^2 \equiv 1 \pmod 2$.

Before we proceed to derive an expression for all the fundamental solutions of (1), we need to prove

LEMMA 2. *If $r^2 = st$ and $(s, t) = 1$, then both s and t are squares.*

PROOF. Write out the prime-power decompositions of s and t:

$$s = p_1^{e_1} p_2^{e_2} \cdots p_k^{e_k}$$
$$t = q_1^{f_1} q_2^{f_2} \cdots q_j^{f_j}.$$

From $(s, t) = 1$, it follows that no prime appears in both decompositions.

Because of the unique factorization theorem, the prime-power decomposition of r^2 can be written

$$r^2 = st = p_1^{e_1} p_2^{e_2} \cdots p_k^{e_k} q_1^{f_1} q_2^{f_2} \cdots q_j^{f_j},$$

and the p's and q's are distinct primes. Since r^2 is a square, all of the exponents $e_1, e_2, \ldots, e_k, f_1, f_2, \ldots, f_j$ are even. Thus s and t are squares.

Exercise 3 (optional). Prove Lemma 2 by induction on r as follows. The lemma is trivially true for $r = 1$ and $r = 2$. Suppose that it is true for $r \leq n - 1$. Note that n has a prime divisor p, and $p \mid s$ or $p \mid t$, but not both. Also, $p^2 \mid n^2$. Apply the induction assumption to n/p^2.

The next lemma gives a condition that fundamental solutions of (1) must satisfy.

LEMMA 3. *Suppose that a, b, c is a fundamental solution of $x^2 + y^2 = z^2$, and suppose that a is even. Then there are positive integers m and n with $m > n$, $(m, n) = 1$ and $m \not\equiv n$ (mod 2) such that*

$$a = 2mn,$$
$$b = m^2 - n^2,$$
$$c = m^2 + n^2.$$

(Note that we lose no generality in assuming that a is even. Lemma 1 tells us that exactly one of a and b is even, so we may as well call the even member of the pair a, b by the name of a.)

PROOF. Since a is even, $a = 2r$ for some r. So, $a^2 = 4r^2$; from $a^2 = c^2 - b^2$ follows

$$(2) \qquad\qquad 4r^2 = (c + b)(c - b).$$

We know that b is odd, and from the Corollary to Lemma 1, we know that c is odd too. Thus $c + b$ and $c - b$ are both even. Thus we can put

$$(3) \qquad\qquad c + b = 2s \quad \text{and} \quad c - b = 2t.$$

Then

$$(4) \qquad\qquad c = s + t \quad \text{and} \quad b = s - t.$$

Substituting (3) into (2), we get $4r^2 = (2s)(2t)$, or

$$r^2 = st.$$

If s and t are relatively prime, we can apply Lemma 2 and conclude that s and t are both squares. In fact, s and t are relatively prime, as we now show. Suppose that $d \mid s$ and $d \mid t$. From (4) it follows that $d \mid b$ and $d \mid c$.

But from Exercise 1, we know that b and c are relatively prime. Hence $d = \pm 1$ and $(s, t) = 1$. Lemma 2 says that

$$s = m^2 \quad \text{and} \quad t = n^2$$

for some integers m and n, which we may assume to be positive. Thus

$$a^2 = 4r^2 = 4st = 4m^2n^2$$

or $a = 2mn$. From (4),

$$c = s + t = m^2 + n^2,$$
$$b = s - t = m^2 - n^2.$$

Having established the last three equations, we need now only establish that $m > n$, $(m, n) = 1$, and $m \not\equiv n \pmod 2$ to complete the proof. The inequality follows because b is part of a fundamental solution and hence positive.

Exercise 4. Suppose that $d \mid m$ and $d \mid n$. Show that $d \mid a$ and $d \mid b$. Conclude that $(m, n) = 1$.

Exercise 5. Suppose that $m \equiv n \equiv 0 \pmod 2$. Show that a and b are both even, which is impossible.

Exercise 6. Suppose that $m \equiv n \equiv 1 \pmod 2$. Show again that a and b are both even.

Exercises 4, 5, and 6 have completed the proof of Lemma 3.

We have shown that if a, b, c is a fundamental solution, then $a, b,$ and c satisfy the conditions of Lemma 3. It is also true that if a, b, c satisfy the conditions of Lemma 3, then a, b, c is a fundamental solution of $x^2 + y^2 = z^2$. We prove this in

LEMMA 4. *If*

$$a = 2mn,$$
$$b = m^2 - n^2,$$
$$c = m^2 + n^2,$$

then a, b, c is a solution of $x^2 + y^2 = z^2$. If in addition, $m > n$, m and n are positive, $(m, n) = 1$, and $m \not\equiv n \pmod 2$, then a, b, c is a fundamental solution.

PROOF. To see that a, b, c is a solution is a matter of computation:

$$a^2 + b^2 = (2mn)^2 + (m^2 - n^2)^2$$
$$= 4m^2n^2 + m^4 - 2m^2n^2 + n^4 = m^4 + 2m^2n^2 + n^4$$
$$= (m^2 + n^2)^2 = c^2.$$

It remains to show that $(m, n) = 1$ and $m \not\equiv n \pmod 2$ imply that $(a, b) =$

1. Suppose that p is an odd prime such that $p \mid a$ and $p \mid b$. From $c^2 = a^2 + b^2$ it follows that $p \mid c$. From $p \mid b$ and $p \mid c$ it follows that $p \mid (b + c)$ and $p \mid (b - c)$. But

$$b + c = 2m^2 \quad \text{and} \quad b - c = -2n^2.$$

So, $p \mid 2m^2$ and $p \mid 2n^2$. Since p is odd, this implies that $p \mid m^2$ and $p \mid n^2$, and hence that $p \mid m$ and $p \mid n$. Since m and n are relatively prime, this is impossible. The only way in which a and b could fail to be relatively prime is for both to be divisible by 2. But b is odd because $b = m^2 - n^2$ and one of m, n is even and the other is odd. Thus $(a, b) = 1$, and because $m > n$, b is positive. Because m and n are positive, a is positive. Thus a, b, c is a fundamental solution.

We restate Lemmas 3 and 4 as

THEOREM 1. *All solutions* $x = a$, $y = b$, $z = c$ *to* $x^2 + y^2 = z^2$, *where* a, b, c *are positive and have no common factor and a is even, are given by*

$$a = 2mn,$$
$$b = m^2 - n^2,$$
$$c = m^2 + n^2,$$

where m and n are any relatively prime integers, not both odd, and $m > n$.

Here is a table of some fundamental Pythagorean triangles with small sides:

m	n	a	b	c	a^2	b^2	c^2
2	1	4	3	5	16	9	25
3	2	12	5	13	144	25	169
4	1	8	15	17	64	225	289
4	3	24	7	25	576	49	625
5	2	20	21	29	400	441	841
5	4	40	9	41	1600	81	1681
6	1	12	35	37	144	1225	1369
7	2	28	45	53	784	2025	2809

Problems

1. Besides those in the above table, how many Pythagorean triangles are there with hypotenuse less than 50?

2. Find a Pythagorean triangle all of whose sides are greater than 50.

3. If $(a, b) = d$ and $a^2 + b^2 = c^2$, show that $(a, c) = (b, c) = d$.

4. (a) If m and n generate a, b, c as in Theorem 1, show that $(b + c)/a = m/n$.
   (b) Which m, n generate $72^2 + 65^2 = 97^2$?

5. If $(a, b) = 1$ and $ab = c^n$, does it follow that a and b are nth powers?

6. If $(a, b) = d$ and $ab = c^n$, show that a/d and b/d are nth powers.

7. Prove that if the sum of two consecutive integers is a square, then the smaller is a leg, and the larger a hypotenuse, of a Pythagorean triangle.

8. Bhascara (ca. 1150) found a right triangle whose area is numerically equal to the length of its hypotenuse. Show that this cannot happen if the triangle has integer sides.

9. In all of the Pythagorean triangles in the table in the text, one side is a multiple of five. Is this true for all Pythagorean triangles?

10. Show that 12 divides the product of the legs of a Pythagorean triangle.

11. Show that 60 divides the product of the sides of a Pythagorean triangle.

12. Here is a quadrilateral, not a parallelogram, with integer sides and integer area:

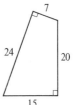

(a) What is its area?
(b) Such quadrilaterals are not common; can you find another?
(c) Could you find 1,000,000 more?

13. Let the generators of a Pythagorean triangle be consecutive triangular numbers. Show that one side of the triangle generated is a cube.

14. Prove that 3, 4, 5 is the only solution of $x^2 + y^2 = z^2$ in consecutive positive integers.

15. Show that the only Pythagorean triangles with sides in arithmetic progression are those with sides $3n, 4n, 5n, n = 1, 2, 3, \ldots$.

16. $3^2 + 4^2 = 5^2.\ 5^2 + 12^2 = 13^3.\ 7^2 + 24^2 = 25^2.\ 9^2 + 40^2 = 41^2.$

(a) Guess a theorem.
(b) Prove that the numbers in (a) give the *only* Pythagorean triangles with consecutive integers for one leg and a hypotenuse.

17. (a) Look in the table in the text and find two Pythagorean triangles with the same area.
(b) Can you find two others with the same area?
(c) Prove that two Pythagorean triangles with the same area and equal hypotenuses are congruent.

18. Show that $n^2 + (n + 1)^2 = 2m^2$ is impossible.

19. Show that $n^2 + (n + 1)^2 = km^2$ is possible only when -1 is a quadratic residue (mod k).

20. Note that $4^2 - 3^2 = 7,\ 12^2 - 5^2 = 7 \cdot 17$, and $8^2 - 15^2 = -7 \cdot 23$. Show that $a^2 + b^2 = c^2$ and $(7, abc) = 1$ imply $7 \mid (a^2 - b^2)$.

21. Although 9 is not the sum of two positive integer squares, it is the sum of two positive *rational* squares. Find them.

22. (a) Given a, how would you find b such that $a^2 + b^2$ is a square?
(b) Carry out such a procedure for $a = 13$ and $a = 14$.

23. Find a Pythagorean triangle whose area is numerically equal to its perimeter.

24. (a) $3^2 + 4^2 = 5^2$. $20^2 + 21^2 = 29^2$. $119^2 + 120^2 = 169^2$. To find another such relation, show that if $a^2 + (a + 1)^2 = c^2$, then

$$(3a + 2c + 1)^2 + (3a + 2c + 2)^2 = (4a + 3c + 2)^2.$$

(b) Find another relation like that in (a).

(c) If $a^2 + (a + 1)^2 = c^2$, let $u = c - a - 1$ and $v = (2a + 1 - c)/2$. Show that v is an integer and that $u(u + 1)/2 = v^2$. (This shows that there are infinitely many square triangular numbers.)

Infinite Descent and Fermat's Conjecture

In the section on Pythagorean triangles, we found all of the solutions in integers of $x^2 + y^2 = z^2$. After disposing of that problem, it would be natural to try the same ideas on an equation of one higher degree, $x^3 + y^3 = z^3$. The same ideas would not work, nor would any others; there is no solution in integers of $x^3 + y^3 = z^3$. (There is one exception—a solution in which one of the variables is zero. We will call such a solution a *trivial* solution, treating it with the contempt it deserves. When we say "solution" in this section, we will mean "nontrivial solution".) In fact, no one knows any solution in integers to any of the equations $x^n + y^n = z^n$ for $n \geq 3$. Fermat thought he had a proof that $x^n + y^n = z^n$ has no nontrivial solutions when $n \geq 3$; he wrote a note in the margin of his copy of the works of Diophantus saying that he had a proof, but that the margin was too small to contain it. It is almost certain that he was mistaken. But of course we cannot be certain. He may have had a proof. Or, he may have realized how deep the proof must lie and wrote the comment to keep future generations of mathematicians at work. Are Frenchmen practical jokers?

The statement "$x^n + y^n = z^n$ has no nontrivial solutions if $n \geq 3$" is often called "Fermat's Last Theorem"—to distinguish it from the theorem that bears his name (see Section 6), but a better name would be *Fermat's Conjecture*. An enormous amount of work has been devoted to it, but it is still not settled one way or the other. The conjecture is known to be true for $n < 25{,}000$ and for many larger values of n, but this is far from a proof. Before the First World War, there was a large prize offered in Germany for a correct proof, and many amateurs offered attempted solutions. It is said that the great

number theorist Landau had post cards printed which read, "Dear Sir or Madam: Your attempted proof of Fermat's Theorem has been received and is herewith returned. The first mistake is on page ——, line ——.";
Landau would give them to his students to fill in the missing numbers. Many powerful mathematicians have worked on the problem. It may be that Fermat's conjecture is forever undecidable one way or the other, for there may exist a solution of $x^n + y^n = z^n$ in numbers so large that no one could ever find them. There are, after all, integers so big that the world could not hold them if they were written out. If they take up that much room, can they fit into the head of man?

The object of this section is to show that Fermat's conjecture is true for $n = 4$, and in so doing, to illustrate Fermat's method of infinite descent. (You might wonder why we avoid considering the case $n = 3$. It turns out to be harder to show that $x^3 + y^3 = z^3$ has no solutions than it is to show that $x^4 + y^4 = z^4$ has none, though it is also possible to apply the method of infinite descent when $n = 3$. See, for example, *Diophantine Analysis*, by R. D. Carmichael (Dover, 1959).) We will prove

THEOREM 1. *There are no nontrivial solutions of*

$$x^4 + y^4 = z^2.$$

Note that this implies that there are no solutions of $x^4 + y^4 = z^4$; if a, b, c were a solution to that equation, we would have $a^4 + b^4 = (c^2)^2$, contrary to Theorem 1.

PROOF OF THE THEOREM. We will apply Fermat's method of infinite descent. Consider the set of nontrivial solutions of $x^4 + y^4 = z^2$. We want to show that it is empty. We will suppose that it is not empty and deduce a contradiction. Among the set of nontrivial solutions, there is one with a smallest value of z^2. Let c^2 denote this value of z^2. There may be several solutions with this same value of z; if there are, we will pick any one of them—it makes no difference which. Call the solution that we pick a, b, c. The idea of the proof is to construct numbers r, s, t that also satisfy $x^4 + y^4 = z^2$, with $t^2 < c^2$. Since c^2 was chosen as small as possible, it follows that the assumption that the set of solutions was not empty was wrong. Hence there are no nontrivial solutions. This method is no mere trick, but is quite natural. It is very possible that Fermat one day set himself to the task of finding solutions of $x^4 + y^4 = z^4$; he may have applied various devices to reduce the equation in the same way as we reduced the equation $x^2 + y^2 = z^2$; and maybe he was surprised when the result of his efforts was another equation of the same form, but with smaller numbers—surprised and pleased too, because this allowed him to conclude that if there was a solution, then there was another smaller solution, and then another and another and another: an infinitely descending chain of solutions. But since we may assume that x, y, and z are positive, this is impossible. (On the other hand, Fermat may have sat down and thought, "I will now apply my method of infinite descent to $x^4 + y^4 = z^4$"; history is silent on the subject.)

We suppose that we have a nontrivial solution a, b, c with c^2 as small as possible. We can suppose that a and b are relatively prime. Suppose not. Then there is a prime p such that $p \mid a$ and $p \mid b$, and hence $p^2 \mid c$. Thus $(a/p)^4 + (b/p)^4 = (c/p^2)^2$ provides a solution to $x^4 + y^4 = z^2$ with a smaller value of z^2 than c^2, and we have supposed this to be impossible.

Exercise 1. Show that a and b cannot both be odd. (Consider $a^4 + b^4 = c^2$ modulo 4.)

Because $(a, b) = 1$, a and b cannot both be even, either. Thus one is even and one is odd. Since $a^4 + b^4 = c^2$ is symmetric in a and b, we can agree to call the even member of the pair a,b by the name of a. But now we have a fundamental solution of $x^2 + y^2 = z^2$ as defined in the last section:

$$(a^2)^2 + (b^2)^2 = c^2,$$

$(a^2, b^2) = 1$, and a^2 is even and b^2 is odd. Hence, by Lemma 3 of the last section, there are integers m and n, relatively prime and not both odd, such that

$$
\begin{aligned}
a^2 &= 2mn, \\
b^2 &= m^2 - n^2, \\
c &= m^2 + n^2.
\end{aligned}
$$

(1)

Exercise 2. Show that n is even. (Suppose that n is odd and m is even, and look at $b^2 = m^2 - n^2$ in the modulus 4. Remember that $x^2 \equiv -1 \pmod 4$ is impossible.)

Because n is even, it follows that m is odd. Put

$$n = 2q.$$

Then from (1), $a^2 = 4mq$, or

(2)
$$\left(\frac{a}{2}\right)^2 = mq.$$

We would like to conclude that m and q are both squares. To do that, we need, according to Lemma 2 of the last section, to show that m and q are relatively prime.

Exercise 3. Show that $(m, q) = 1$. (Suppose not. Then $(m, n) \neq 1$.)

So, there are integers t and v such that

$$m = t^2 \quad \text{and} \quad q = v^2.$$

Exercise 4. Verify that t and v are relatively prime. (Suppose not. Then $(m, q) \neq 1$.)

Exercise 5. Note that t is odd. (Suppose not. Then m is even.)

So far, we have found out a good many things about a, b, and c. We need more yet. We start with the easy observation

$$n^2 + (m^2 - n^2) = m^2.$$

Substituting into this the various facts we know,

$$n = 2q = 2v^2, \quad m^2 - n^2 = b^2, \quad m = t^2,$$

the equation becomes

$$(2v^2)^2 + b^2 = (t^2)^2,$$

so we have another Pythagorean triangle. Is $2v^2$, b, t^2 a fundamental solution? It is if $2v^2$ and b are relatively prime. They are, as we will discover in

Exercise 6. Supply the reasons for the following implications:

> if $p \mid 2v^2$ and $p \mid b$, then $p \mid n$ and $p \mid b$,
> if $p \mid n$ and $p \mid b$, then $p \mid n$ and $p \mid m$,
> if $(m, n) = 1$, then $(2v^2, b) = 1$.

With $2v^2$ even, we have a fundamental solution of $x^2 + y^2 = z^2$, so we need only apply Lemma 3 of the last section to show that there are integers M and N, with $(M, N) = 1$ and $M \not\equiv N \pmod 2$, such that

$$2v^2 = 2MN,$$
$$(3) \qquad b = M^2 - N^2,$$
$$t^2 = M^2 + N^2.$$

Thus we have $v^2 = MN$ and $(M, N) = 1$. The product of two relatively prime integers is a square if and only if both integers are squares (Lemma 2 of the last section), so there are integers r and s such that

$$M = r^2 \quad \text{and} \quad N = s^2.$$

From the third equation in (3), we have

$$t^2 = (r^2)^2 + (s^2)^2,$$

or

$$r^4 + s^4 = t^2.$$

Here is another solution of $x^4 + y^4 = z^2$. It has the property that

$$t^2 = m \leq m^2 < m^2 + n^2 = c \leq c^2,$$

which is impossible, because c^2 was chosen as small as possible. This contradiction proves the theorem.

Exercise 7. Why $m^2 < m^2 + n^2$? Why not just $m^2 \leq m^2 + n^2$?

Here is another, slightly different example of the method of infinite descent, considerably less important than Theorem 1. Suppose that we want to find integers a, b, and c such that

$$\begin{vmatrix} 1 & a & b \\ a & 1 & c \\ b & c & 1 \end{vmatrix} = 1.$$

If we expand the determinant, this is the same as asking for integers a, b, and c that satisfy

(4) $$x^2 + y^2 + z^2 = 2xyz.$$

Exercise 8. Show that it is impossible for all three of a, b, c to be odd. Show that it is impossible for two of them to be even and the other odd.

It is also impossible for two of them to be odd and the other even. (If this were the case, (4), taken modulo 4, would say that $2 \equiv 0 \pmod 4$.) It follows that if a, b, c satisfy (4), then they are all even. So, $a/2$, $b/2$, and $c/2$ are all integers, and dividing (4) by 4 gives

(5) $$(a/2)^2 + (b/2)^2 + (c/2)^2 = 4(a/2)(b/2)(c/2).$$

The same arguments used to show that a, b, c were all even will also show that $a/2$, $b/2$, $c/2$ are all even. Hence $a/4$, $b/4$, and $c/4$ are integers, and dividing (5) by 4 gives

$$(a/4)^2 + (b/4)^2 + (c/4)^2 = 8(a/4)(b/4)(c/4).$$

The same arguments once again show that $a/8$, $b/8$, and $c/8$ are integers. And so on: we find that $a/2^n$, $b/2^n$, and $c/2^n$ are integers for all positive integers n. The only integers that satisfy that condition are $a = b = c = 0$, and thus this is the only solution of (4).

Problems

1. Show that $x^{4n} + y^{4n} = z^{4n}$, $n = 1, 2, \ldots$, has no nontrivial solutions.

2. Suppose that we can show that $x^p + y^p = z^p$ has no nontrivial solutions for any odd prime p. Conclude from this and Problem 1 that $x^n + y^n = z^n$ has no solutions for any $n \geq 3$.

3. Show that $x^p + y^p = z^p$ implies $p \mid (x + y - z)$.

4. Show that $x^{p-1} + y^{p-1} = z^{p-1}$ has no nontrivial solutions unless $p \mid xyz$.

5. Are there any fundamental solutions of $x^2 + (x + 2)^2 = y^2$?

6. Consider $x^2 + y^2 + z^2 = kxyz$. For what values of k does the argument carried out in the text show that the only solution to this equation is $x = y = z = 0$?

7. Show that there are no nontrivial solutions to

$$\text{(a) } x^2 + y^2 = x^2 y^2, \qquad \text{(b) } x^2 + y^2 + z^2 = x^2 y^2.$$

8. Show that there are infinitely many nontrivial solutions of

$$x^n + y^n = z^{n+1}$$

for any $n \geq 1$, namely those given by

$$x = (ac)^{rn}, \qquad y = (bc)^{rn}, \qquad z = c^s,$$

where

$$c = a^{rn^2} + b^{rn^2},$$

a and b are arbitrary, and r and s are chosen to satisfy

$$rn^2 + 1 = (n + 1)s.$$

Does the last equation have infinitely many solutions in positive integers r, s?

9. Find a solution of $x^4 + y^4 = z^5$.

10. Show that $x^n + y^n = z^n$ has no solutions with both x and y less than n for any positive integer n.

11. Find solutions to $x^n + y^n = z^{n-1}$.

12. Show that $x^n + y^n = z^m$ has nontrivial solutions if $(n, m) = 1$.

Sums of
Two Squares

Among the integers from 1 to 99, the following 57 are not representable as the sum of two squares of integers.

3	6	7	11	12	14	15	19	21	22	23
24	27	28	30	31	33	35	38	39	42	43
44	46	47	48	51	54	55	56	57	59	60
62	63	66	67	69	70	71	75	76	77	78
79	83	84	86	87	88	91	92	93	94	95
96	99									

But the remaining 42 are.

1	2	4	5	8	9	10	13	16	17	18
20	25	26	29	32	34	36	37	40	41	45
49	50	52	53	58	61	64	65	68	72	73
74	80	81	82	85	89	90	97	98		

It would be a good exercise of your inductive powers to look at these lists, and in the spirit of the scientific method, try to formulate a hypothesis that would explain the presence of a number on its list and which could be used to predict results for numbers greater than 99. There is a fairly simple property (other than not being representable as a sum of two squares) that the numbers in the first list share. As an aid to looking in the right direction, we prove

LEMMA 1. *If* $n \equiv 3 \ (mod \ 4)$, *then* $n = x^2 + y^2$ *is impossible.*

PROOF. Since $x^2 \equiv 0$ or $1 \pmod 4$ for all integers x, it follows that

$$x^2 + y^2 \equiv 0, 1, \text{ or } 2 \pmod 4$$

for any x and y, so $x^2 + y^2 \equiv 3$ (mod 4) is impossible. This lemma accounts for the presence of 25 of the 57 numbers in the first list.

LEMMA 2. *If n is representable, then so is $k^2 n$ for any k.*

PROOF. If $n = x^2 + y^2$, then $k^2 n = (kx)^2 + (ky)^2$.
(Instead of "n is representable as the sum of two squares" we will sometimes say "n is representable" for short.)

If you have given up, here is the answer:

THEOREM 1. *n cannot be written as the sum of two squares if and only if the prime-power decomposition of n contains a prime congruent to 3 (mod 4) to an odd power.*

PROOF (of the "if" part). Suppose that p is a prime, $p \equiv 3$ (mod 4), which appears in the prime-power decomposition of n to an odd power. That is, for some $e \geq 0$, $p^{2e+1} \mid n$ and $p^{2e+2} \nmid n$. Suppose that $n = x^2 + y^2$ for some x and y. We will deduce a contradiction—namely that -1 is a quadratic residue (mod p), which is impossible, since $p \equiv 3$ (mod 4). Let $d = (x, y)$, $x_1 = x/d$, $y_1 = y/d$, and $n_1 = n/d$. Then

$$(1) \qquad\qquad x_1^2 + y_1^2 = n_1 \qquad \text{and} \qquad (x_1, y_1) = 1.$$

If p^f is the highest power of p that divides d, then n_1 is divisible by $p^{2e-2f+1}$, and this exponent, being odd and nonnegative, is at least one. Thus $p \mid n_1$, and since $(x_1, y_1) = 1$, we have $p \nmid x_1$. Hence there is a number u such that

$$x_1 u \equiv y_1 \pmod{p}.$$

Thus

$$(2) \qquad 0 \equiv n_1 \equiv x_1^2 + y_1^2 \equiv x_1^2 + (ux_1)^2 \equiv x_1^2(1 + u^2) \pmod{p}.$$

Since $(x_1, p) = 1$, x_1 may be cancelled in (2) to give

$$1 + u^2 \equiv 0 \pmod{p}.$$

This says that -1 is a quadratic residue (mod p), which is impossible. Hence $n = x^2 + y^2$ is impossible, and the easy part of the theorem is proved.

The rest of the section will be devoted to proving the rest of Theorem 1 (the "only if" part). We will need two lemmas.

LEMMA 3. *$(a^2 + b^2)(c^2 + d^2) = (ac + bd)^2 + (ad - bc)^2$ for any integers a, b, c, and d.*

PROOF. Multiply it out.

The result shows that if two numbers are representable, then so is their product.

LEMMA 4. *Any integer n can be written in the form*

$$n = k^2 p_1 p_2 \cdots p_r,$$

where k is an integer and the p's are different primes.

Exercise 1. Convince yourself that this is so by considering the prime-power decomposition of n.

As an example of the application of Lemmas 2 and 3, we can get a representation for $260 = 2^2 \cdot 5 \cdot 13$ from the representations

$$5 = 2^2 + 1^2 \quad \text{and} \quad 13 = 3^2 + 2^2.$$

From Lemma 3,

$$65 = 5 \cdot 13 = (2^2 + 1^2)(3^2 + 2^2)$$
$$= (2 \cdot 3 + 1 \cdot 2)^2 + (2 \cdot 2 - 1 \cdot 3)^2 = 8^2 + 1^2.$$

Hence

$$260 = 2^2 \cdot 65 = (8 \cdot 2)^2 + (1 \cdot 2)^2 = 16^2 + 2^2.$$

Exercise 2. Write 325 as a sum of two squares.

Exercise 3. If the prime-power decomposition of n contains no prime p, $p \equiv 3$ (mod 4), to an odd power, then note that

$$n = k^2 p_1 p_2 \cdots p_r \quad \text{or} \quad n = 2k^2 p_1 p_2 \cdots p_r$$

for some k and r, where each p is congruent to 1 (mod 4).

Exercise 3 and Lemmas 2 and 3 imply that to prove the rest of Theorem 1, it is sufficient to prove

THEOREM 2. *Every prime congruent to 1 (mod 4) can be written as a sum of two squares.*

PROOF. The idea of the proof is this: if $p \equiv 1$ (mod 4), then we show that there are nonzero integers x and y such that

$$x^2 + y^2 = kp$$

for some integer k, $k \geq 1$. If $k > 1$, we then construct from x and y new integers x_1 and y_1 such that

$$x_1^2 + y_1^2 = k_1 p$$

for some k_1, with $k_1 < k$. This is the step that proves the theorem, because if $k_1 > 1$, we repeat the process to get integers x_2 and y_2 such that $x_2^2 + y_2^2 = k_2 p$ with $k_2 < k_1$. If we keep on, we will get a decreasing sequence of positive integers, $k > k_1 > k_2 > \cdots$, which must eventually reach 1. When it does, we have a representation of p as a sum of two squares.

First we show that we can find nonzero x and y such that $x^2 + y^2 = kp$ for some k, $k \geq 1$. Since $p \equiv 1 \pmod 4$, we know that -1 is a quadratic residue $\pmod p$; hence there is an integer u such that $u^2 \equiv -1 \pmod p$. That is, $p \mid (u^2 + 1)$, or

$$u^2 + 1 = kp$$

for some k, $k \geq 1$. Hence $x^2 + y^2 = kp$ always has a solution for some k, $k \geq 1$; in fact, we can take $y = 1$. For example, if $p = 17$, we have $4^2 + 1^2 = 1 \cdot 17$; if $p = 29$, $12^2 + 1^2 = 5 \cdot 29$. The number u can be found by trial (we can write $kp - 1$ for $k = 1, 2, \ldots$, and continue until we come to a square), or we can use the fact that

$$\left(\left(\frac{p-1}{2} \right)! \right)^2 \equiv -1 \pmod p.$$

The last congruence (which follows from Wilson's Theorem) gives a construction for u: a long one, perhaps, but one which guarantees a result.

We now show how to construct x_1 and y_1. Define r and s by

$$(3) \quad r \equiv x \pmod k, \qquad s \equiv y \pmod k, \qquad -\frac{k}{2} < r \leq \frac{k}{2}, \quad -\frac{k}{2} < s \leq \frac{k}{2}.$$

From (3),

$$r^2 + s^2 \equiv x^2 + y^2 \pmod k.$$

But we had chosen x and y such that $x^2 + y^2 = kp$. Hence

$$r^2 + s^2 \equiv 0 \pmod k,$$

or

$$(4) \qquad r^2 + s^2 = k_1 k$$

for some k_1. It follows from (4) that

$$(r^2 + s^2)(x^2 + y^2) = (k_1 k)(kp) = k_1 k^2 p.$$

From Lemma 3, however, we have

$$(r^2 + s^2)(x^2 + y^2) = (rx + sy)^2 + (ry - sx)^2.$$

Thus

$$(5) \qquad (rx + sy)^2 + (ry - sx)^2 = k_1 k^2 p.$$

Note that from (3),

$$rx + sy \equiv r^2 + s^2 \equiv 0 \pmod{k}$$

and

$$ry - sx \equiv rs - sr \equiv 0 \pmod{k}.$$

Thus k^2 divides each term on the left-hand side of (5); dividing (5) by k^2, we get

$$\left(\frac{rx + sy}{k}\right)^2 + \left(\frac{ry - sx}{k}\right)^2 = k_1 p,$$

an equation in integers. Let $x_1 = (rx + sy)/k$ and $y_1 = (ry - sx)/k$. Then $x_1^2 + y_1^2 = k_1 p$, and the theorem will be proved when we show that $k_1 < k$. The inequalities in (3) give

$$r^2 + s^2 \le (k/2)^2 + (k/2)^2 = k^2/2.$$

But

$$r^2 + s^2 = k_1 k.$$

Thus $k_1 k \le k^2/2$ or $k_1 \le k/2$. Hence $k_1 < k$, and the theorem is proved. (Note that $k_1 \ge 1$: if $k_1 = 0$, then from the last equation, $r = s = 0$.)

Exercise 4. Why is $r = s = 0$ impossible?

Let us take an example. Starting with $12^2 + 1^2 = 5 \cdot 29$, we will carry through the calculations of the theorem to get a representation of 29 as a sum of two squares. We have $x = 12$, $y = 1$, and $k = 5$. We then have $r \equiv 12$ (mod 5) and $s \equiv 1$ (mod 5); choosing r and s in the proper range, we get $r = 2$ and $s = 1$. Then

$$5^2 \cdot 29 = (2^2 + 1^2)(12^2 + 1^2) = (2 \cdot 12 + 1 \cdot 1)^2 + (2 \cdot 1 - 1 \cdot 12)^2$$
$$= 25^2 + 10^2.$$

Dividing by $k^2 = 25$ gives $29 = 5^2 + 2^2$, the desired representation.

Exercise 5. Try the calculation for $23^2 + 1^2 = 10 \cdot 53$.

We will end with some remarks on diophantine equations closely related to the sum of two squares. We have completely solved the problem of representing integers as the sum of two squares: we know which integers can be so represented, and Theorem 2, combined with earlier lemmas, gives a method for actually calculating the representation. It is natural now to wonder about the representations of integers as the sum of *three* squares. We would expect that more integers can be represented when we have an extra square to add, and this is the case. It is a fact that n can be written as a sum of three squares unless $n = 4^e(8k + 7)$ for some integers e and k. So, the numbers smaller than 100 which cannot be written as the sum of three squares are

$$7,\ 15,\ 23,\ 28,\ 31,\ 39,\ 47,\ 55,\ 60,\ 63,\ 71,\ 79,\ 87,\ 92,\ 95:$$

a total of 15, as against 57 when only two squares were allowed. There are even fewer exceptions if we look at sums of *four* squares; in fact, there are none at all. Every integer can be written as a sum of four squares, as we will show in the next section.

That would seem to settle the squares. (That is, as far as the mere representation is concerned. Of course, there are many, many other questions that can be asked, and some that can be answered. For example, how many representations does an integer have as a sum of two squares? What is the sum of the number of solutions of $x^2 + y^2 = n^2$ for $n = 1, 2, \ldots, N$? And so on: as soon as one question is settled, others crowd in to take its place.) What about cubes? It is true that every integer can be written as a sum of nine cubes. No one knows what the corresponding number is for the sum of fourth powers (though it is known that every integer greater than $10^{10^{89}}$ is a sum of 19 fourth powers), and it is unknown how many fifth powers are needed to represent every number. But the answer is known for sixth, seventh, and almost all higher powers—for example, 73 sixth powers will do. Let $g(k)$ be the least value of s such that every integer can be written as a sum of no more than s kth powers. The problem of finding $g(k)$ is called *Waring's Problem*. In 1770, Waring wrote that every integer was the sum of 4 squares, 9 cubes, 19 fourth powers, and so on. He was just guessing; it was not until 1909 that it was proved that the number $g(k)$ exists for each k, and even then almost nothing was known about its size for large k. Work since then has shown that for $6 \leq k \leq 200{,}000$, and for all "sufficiently large" k,

$$g(k) = 2^k + \left[\left(\frac{3}{2}\right)^k\right] - 2,$$

and it is strongly suspected that this formula is true for all $k \geq 6$. (The notation $[(3/2)^k]$ denotes the integral part of $(3/2)^k$: see Appendix B, p. 196.) Of more interest than $g(k)$ is $G(k)$, the least value of s such that every *sufficiently large* integer can be written as a sum of no more than s kth powers. $G(2) = 4$ and $G(4) = 16$, but G is not known for any larger values of k. The most that is known in general is that

$$k + 1 \leq G(k) \leq k(3 \log k + 11).$$

The right-hand inequality is extremely difficult and complicated to prove, and its accomplishment (by Vinogradov, building on previous results of Hardy and Littlewood) represents one of the pinnacles of number theory.

Another famous problem about representing integers as sums of certain other integers is *Goldbach's Conjecture*. In 1742, Goldbach noted that

$$4 = 2 + 2, \qquad 6 = 3 + 3, \qquad 8 = 5 + 3, \qquad 10 = 5 + 5,$$
$$12 = 7 + 5, \qquad 14 = 7 + 7, \qquad \dots, \qquad 100 = 97 + 3,$$

and guessed that every even integer greater than two could be written as the sum of two primes. He asked Euler if he could prove it. Euler failed, and no one since has succeeded. The most we can say is that every even integer greater than two can be written as a sum of no more than 20,000,000,000 primes. Using different methods, Vinogradov has shown that every even integer which is sufficiently large is a sum of no more than 4 primes. And it was later shown that "sufficiently large" could be replaced with "larger than $e^{e^{16.038}}$." (Here, e does not represent an integer, but rather the base of natural logarithms, 2.718. . . .) In another direction, it was announced (1966) that every sufficiently large even integer is the sum of a prime and an integer that has no more than two different prime factors. This is as close as the conjecture has come to being settled. (For these results, see *Mathematical Reviews*, vol. 34 (1967), reviews 5784 and 7483.)

Problems

1. Determine which of the following integers can be written as a sum of two squares, and for those that can, find a representation.

$$150, 151, 152, 153, 154.$$

2. Which of 1965, 1966, 1967, 1968, 1969, 1970 can be written as a sum of two squares?

3. Represent 10,045, 10,048, and 10,049 as sums of two squares.

4. Prove, directly and without using Theorem 1, that if $7 \mid n$ and $7^2 \nmid n$, then $n = x^2 + y^2$ is impossible.

5. Show that if $n \equiv 3$ or 6 (mod 9), then n is not representable as a sum of two squares.

6. Is it true that if $5 \mid n$ and $5^2 \nmid n$, then $n = x^2 + y^2$ is impossible?

7. Prove that if m and n are sums of two squares and $m \mid n$, then n/m is a sum of two squares.

8. Fermat said

   $2n + 1$ is the sum of two squares when and only when (i) n is even, and (ii) $2n + 1$, when divided by the largest square entering into it as a factor must not be divisible by a prime $4k - 1$.

   Show that this is equivalent to Theorem 1.

9. Girard (1632) said that the numbers representable as the sum of two squares comprise every square, every prime $4k + 1$, a product of such numbers, and the double of any of the preceding. Show that this is equivalent to Theorem 1.

10. Which integers in the range $100 \le n \le 150$ cannot be written as a sum of three squares?

11. Show that $4 \mid (x^2 + y^2 + z^2)$ implies that x, y, and z are even.

12. Verify that if $n \equiv 7 \pmod 8$, then n cannot be written as a sum of three squares.

13. From Problems 11 and 12, show that $n = 4^e(8k + 7)$ for some nonnegative e and k implies that $n = x^2 + y^2 + z^2$ is impossible.

14. A mathematician said in 1621 that $3n + 1$ is not the sum of three squares if $n = 8k + 2$ or $32k + 9$ for some k. Show him to be wrong.

15. Try to mimic the proof of Theorem 2 to prove a generalization: if $(-w/p) = 1$, then there are integers x and y such that $p = x^2 + wy^2$.

16. Show that if n is the sum of two triangular numbers, then $4n + 1$ is a sum of two squares.

17. Which integers can be written as the sum of two squares of *rational* numbers?

Sums of
Four Squares

In this section we will prove that every positive integer can be written as a sum of four squares of integers, some of which may be zero. This theorem is quite old. Diophantus seems to have assumed that every positive integer is a sum of 2, 3, or 4 squares of positive integers, but he never explicitly stated the theorem. The first to do so was Bachet (1621). He verified that it was true for integers up to 325, but he was unable to prove it. Fermat said that he was able to prove it using his method of descent; as usual, he gave no details. In the light of subsequent work on the theorem, we may doubt that Fermat's proof was complete. Descartes said that the theorem was no doubt true, but he judged the proof "so difficult that I dared not undertake to find it." (It is hard to resist the temptation to read "so difficult that I was unable to find it.")

Euler next took up the challenge, first working on the problem in 1730. In 1743 he noted that the product of two sums of four squares is again a sum of four squares, a result fundamental to the proof of the theorem, and indeed, he proved the theorem except for one point. In 1751, still pursuing that point, Euler proved another fundamental result, namely that $1 + x^2 + y^2 \equiv 0$ (mod p) always has a solution for any prime p. But the theorem was still out of his reach. Finally, in 1770, Lagrange, drawing heavily on Euler's ideas, succeeded in constructing a proof. In 1773, Euler (then 66 years old) gave a simpler proof—success after 43 years.

We start by proving Euler's two results.

LEMMA 1. *The product of two sums of four squares is a sum of four squares.*

PROOF. The proof is utterly trivial. Its discovery was quite another matter, as witness the thirteen-year gap between Euler's first attack on the problem and his discovery of the following identity (almost any identity is trivial when written down by someone else):

$$(a^2 + b^2 + c^2 + d^2)(r^2 + s^2 + t^2 + u^2) = (ar + bs + ct + du)^2$$
$$+ (as - br + cu - dt)^2 + (at - bu - cr + ds)^2$$
$$+ (au + bt - cs - dr)^2,$$

which may be verified by multiplication. (Note that the right-hand side, when multiplied out, contains all the terms a^2r^2, a^2s^2, . . . , d^2u^2 that appear on the left-hand side multiplied out. Note also that it is not impossible to see by inspection that all the cross-product terms vanish.)

It follows from Lemma 1 that, in order to show that every positive integer is the sum of four squares, we need only show that every prime is the sum of four squares. For example, from

$$37 = 6^2 + 1^2 + 0^2 + 0^2$$

and

$$57 = 7^2 + 2^2 + 2^2 + 0^2,$$

we get

$$2109 = 57 \cdot 37 = (6 \cdot 7 + 1 \cdot 2 + 0 \cdot 2 + 0 \cdot 0)^2 + (6 \cdot 2 - 1 \cdot 7 + 0 \cdot 0 - 0 \cdot 2)^2$$
$$+ (6 \cdot 2 - 1 \cdot 0 - 0 \cdot 7 + 0 \cdot 2)^2 + (6 \cdot 0 + 1 \cdot 2 - 0 \cdot 2 - 0 \cdot 7)^2$$
$$= 44^2 + 5^2 + 12^2 + 2^2;$$

we can similarly decompose any integer into a product of primes and then get a representation of it as a sum of four squares if we know a representation of each prime as a sum of four squares.

LEMMA 2. *If p is an odd prime, then*

$$1 + x^2 + y^2 \equiv 0 \ (mod \ p)$$

has a solution with $0 < x < p/2$ and $0 < y < p/2$.

PROOF. The numbers in

$$S_1 = \left\{ 0^2, 1^2, 2^2, \ldots, \left(\frac{p-1}{2} \right)^2 \right\}$$

are distinct (mod p). (For from $x^2 \equiv y^2$ (mod p) follows $x \equiv \pm y$ (mod p).) So are the numbers in

$$S_2 = \left\{ -1 - 0^2, -1 - 1^2, -1 - 2^2, \ldots, -1 - \left(\frac{p-1}{2} \right)^2 \right\}$$

distinct (mod p). S_1 and S_2 contain together $(p - 1)/2 + 1 + (p - 1)/2 +$

$1 = p + 1$ numbers. Since there are only p least residues (mod p), we must have one of the numbers in S_1 congruent to one of the numbers in S_2:

$$x^2 \equiv -1 - y^2 \text{ (mod } p)$$

for some x and y, and $0 \leq x \leq (p-1)/2$, $0 \leq y \leq (p-1)/2$.

To show that every positive integer is the sum of four squares, we use the same method of proof as was used to prove the theorem on integers that are the sum of two squares. We express some multiple of p as a sum of four squares and then construct a smaller multiple of p also the sum of four squares. Repeating the process often enough will give p as a sum of four squares, and that is all we need. Since $2 = 1^2 + 1^2 + 0^2 + 0^2$, the case $p = 2$ is settled, and we can assume hereafter that p is an odd prime.

LEMMA 3. *For every odd prime p, there is an odd integer m, $m < p$, such that*

$$mp = x^2 + y^2 + z^2 + w^2$$

has a solution.

PROOF. From Lemma 2 we know that there are x and y such that

$$kp = x^2 + y^2 + 1^2 + 0^2$$

for some k. Since $0 < x < p/2$ and $0 < y < p/2$, we have

$$kp = x^2 + y^2 + 1 < p^2/4 + p^2/4 + 1 < p^2,$$

so $k < p$. It remains to show that we can suppose k odd. Suppose that

$$kp = x^2 + y^2 + z^2 + w^2$$

and k is even. Then x, y, z, w are all odd, all even, or two are odd and two are even. In any event, we can rearrange the terms so that

$$x \equiv y \text{ (mod 2)} \qquad \text{and} \qquad z \equiv w \text{ (mod 2)}.$$

Hence

$$\frac{kp}{2} = \left(\frac{x-y}{2}\right)^2 + \left(\frac{x+y}{2}\right)^2 + \left(\frac{z-w}{2}\right)^2 + \left(\frac{z+w}{2}\right)^2.$$

If $k/2$ is even, we can repeat the process and express $(k/4)p$ as a sum of four squares. Since $k \neq 0$, eventually, we will have an odd multiple of p written as a sum of four squares.

Exercise 1. From

$$12 \cdot 17 = 204 = 14^2 + 2^2 + 2^2 + 0^2$$

and

$$12 \cdot 17 = 204 = 13^2 + 5^2 + 3^2 + 1^2,$$

find representations of $3 \cdot 17$ as sums of four squares.

LEMMA 4. *If m and p are odd, $1 < m < p$, and*

$$mp = x^2 + y^2 + z^2 + w^2,$$

then there is a positive integer m_1 with $m_1 < m$ such that

$$m_1 p = x_1^2 + y_1^2 + z_1^2 + w_1^2$$

for some integers x_1, y_1, z_1, w_1.

PROOF. As in the two-squares theorem, we proceed to construct x_1, y_1, z_1, w_1 from x, y, z, w. Choose $A, B, C,$ and D such that

$$A \equiv x, \qquad B \equiv y, \qquad C \equiv z, \qquad D \equiv w \ (\text{mod } m)$$

and such that each lies strictly between $-m/2$ and $m/2$. That is, A is the numerically smallest residue of $x \ (\text{mod } m)$, and similarly for the others. It follows that

$$A^2 + B^2 + C^2 + D^2 \equiv x^2 + y^2 + z^2 + w^2 \ (\text{mod } m),$$

so

$$A^2 + B^2 + C^2 + D^2 = km$$

for some k. Since

$$A^2 + B^2 + C^2 + D^2 < m^2/4 + m^2/4 + m^2/4 + m^2/4 = m^2,$$

we have $0 < k < m$. (If $k = 0$, then m divides each of $x, y, z,$ and w, so $m^2 \mid mp$. This is impossible, because $1 < m < p$.) Thus

$$m^2 kp = (mp)(km) = (x^2 + y^2 + z^2 + w^2)(A^2 + B^2 + C^2 + D^2),$$

and from Lemma 1 we have

$$m^2 kp = (xA + yB + zC + wD)^2 + (xB - yA + zD - wC)^2$$
$$+ (xC - yD - zA + wB)^2 + (xD + yC - zB - wA)^2.$$

The terms in parentheses are divisible by m:

$$xA + yB + zC + wD \equiv x^2 + y^2 + z^2 + w^2 \equiv 0 \ (\text{mod } m),$$
$$xB - yA + zD - wC \equiv xy - yx + zw - wz \equiv 0 \ (\text{mod } m),$$
$$xC - yD - zA + wB \equiv xz - yw - zx + wy \equiv 0 \ (\text{mod } m),$$
$$xD + yC - zB - wA \equiv xw + yz - zy - wx \equiv 0 \ (\text{mod } m).$$

So, if we put

$$x_1 = (xA + yB + zC + wD)/m, \qquad y_1 = (xB - yA + zD - wC)/m,$$
$$z_1 = (xC - yD - zA + wB)/m, \qquad w_1 = (xD + yC - zB - wA)/m,$$

then we have

$$x_1^2 + y_1^2 + z_1^2 + w_1^2 = (m^2 kp)/m^2 = kp.$$

Since $k < m$, the lemma is proved.

THEOREM 1. *Every positive integer can be written as the sum of four integer squares.*

PROOF. Suppose that $n = p_1^{e_1} p_2^{e_2} \ldots p_k^{e_k}$. Starting with Lemma 3, repeated application of Lemma 4 gives a solution of $p_i = x^2 + y^2 + z^2 + w^2$ for each i. From Lemma 1, we can write $p_i^{e_i}$ as a sum of four squares for each i. Applying Lemma 1 again (k times), we can get a representation of $p_1^{e_1} p_2^{e_2} \ldots p_k^{e_k}$ as a sum of four squares.

Problems

1. Express each prime less than or equal to 23 as a sum of four squares.

2. Express

    (a) 121,   (b) 391,   (c) 47321

    as a sum of four squares.

3. Express 5,724,631 as a sum of four squares.

4. From $53 = 7^2 + 2^2 + 0^2 + 0^2$ and Euler's formula (Lemma 1), find a representation of $18179 = 7^3 \cdot 53$ as a sum of four squares.

5. If $8 \mid (x^2 + y^2 + z^2 + w^2)$, show that x, y, z, and w are even.

6. If $n = x^2 + y^2 + z^2 + w^2$, show that by suitable ordering and choice of sign, we can get $x + y + z$ to be a multiple of three.

7. Which numbers among the integers 10, 11, 12, $\ldots$, 20 have a unique representation (unique up to the order of the summands) as a sum of four squares?

8. If $n = x^2 + y^2 + z^2 + w^2$ and x, y, z, w are nonnegative, show that

    $$\min(x, y, z, w) \le n^{1/2}/2 \le \max(x, y, z, w) \le n^{1/2}.$$

9. From $2 \cdot 17 \cdot 1973 = 67082 = 238^2 + 102^2 + 5^2 + 3^2$, find a representation of $17 \cdot 1973$ as a sum of four squares.

10. Fill in any missing details in Euler's original proof of Lemma 2. Suppose that $(-1/p) = 1$. Then there is an integer x such that $1 + x^2 \equiv 0 \pmod{p}$. So,

suppose that $(-1/p) = -1$ and that the lemma is false. Then $1 + 1 - 2 = 0$ shows that $(-2/p) = -1$ and hence that $(2/p) = 1$. Then $1 + 2 - 3 = 0$ shows that $(-3/p) = -1$ and that $(3/p) = 1$. In this way, $1, 2, \ldots, p - 1$ are all quadratic residues (mod p).

11. If $n > 0$ and $8 \mid n$, show that n is not the sum of fewer than eight squares of odd integers.

12. If t is even and x, y, and z have no common factor, show that

$$t^2 = x^2 + y^2 + z^2$$

is impossible.

$$x^2 - Ny^2 = 1$$

The theory of diophantine equations has not been perfected. There are not many theorems that apply to a really wide class of equations. Usually, special equations are attacked with special methods, and what works for $x^3 + 3xy + y^3 = z^3$ may be worthless for solving $x^3 + 4xy + y^3 = z^3$. (On the other hand, the same method might work for both.) The perfect theorem would be one that would let us look at any diophantine equation and decide whether it had solutions. It would be even better if the theorem would let us decide exactly how many solutions there are, and better yet if it would tell us exactly what they are. It is unlikely that this ideal will ever be attained. To illustrate the state which the theory of diophantine equations has reached today, here is one of the most general theorems now known (general in the sense that it applies to a larger class of equations than do other theorems):

THEOREM. *Let*

$$F(x, y) = a_n x^n + a_{n-1} x^{n-1} y + a_{n-2} x^{n-2} y^2 + \cdots + a_0 y^n,$$

and suppose that $F(x, 1) = 0$ has no repeated roots. Then the equation

$$F(x, y) = c,$$

where c is an integer, has only finitely many solutions if $n \geq 3$.

In particular, this theorem says that $ax^n + by^n = c$ has, in general, only finitely many solutions if $n \geq 3$. What if $n < 3$? We have completely analyzed the case $n = 1$ in the section on linear diophantine equations, and, for $n = 2$, we considered a special case in the section on Pythagorean triangles. The general equation when $n = 2$ is too complicated for us to treat here. In this section, we will treat another special case:

$$x^2 - Ny^2 = 1.$$

We will show that if we can find one solution of this equation with $x > 1$, then we can find infinitely many. In fact, if we can find the smallest solution (the one with $x > 1$ as small as possible), then we can find *all* the solutions of the equation.

The equation $x^2 - Ny^2 = 1$ is commonly called Pell's equation. This is the result of a mistake made by Euler, who called it that. Euler was so eminent that everyone has called it that since. But Pell never solved the equation, and there is even doubt that he could have. The mathematical historian E. T. Bell has written, "Pell mathematically was a nonentity and humanly an egregious fraud. It is long past time that his name be dropped from textbooks." Certainly, frauds and nonentities do not deserve the immortality of having an equation called after them. Henceforth, defying common practice, we will call $x^2 - Ny^2 = 1$ *Fermat's equation.*

The equation is always satisfied when $x = \pm 1$ and $y = 0$, whatever the value of N. We will call solutions in which either $x = 0$ or $y = 0$ trivial solutions.

Exercise 1. Find, by trial, a nontrivial solution of $x^2 - 2y^2 = 1$, and one of $x^2 - 3y^2 = 1$.

An efficient way to go about finding a nontrivial solution of $x^2 - Ny^2 = 1$ by trial is to make a table of $1 + Ny^2$ for $y = 1, 2, \ldots$ and inspect it for squares; Table B (p. 216) may be helpful.

In solving $x^2 - Ny^2 = 1$, there is no need to consider negative values of N. If $N \leq -2$, then it is clear that the equation has only the trivial solutions with $y = 0$, because both terms on the left are nonnegative. For $N = -1$, there are also the solutions $x = 0$, $y = \pm 1$. These are trivial too. Besides supposing N to be positive, we can assume that N is not a square. If it is, then $N = m^2$ for some m, and we have

$$1 = x^2 - m^2y^2 = (x - my)(x + my).$$

The product of two integers is 1 only when both are 1 or both are -1, and all the solutions can thus be quickly found by solving pairs of linear equations.

Exercise 2 (optional). Show that $x = \pm 1$, $y = 0$ are the only solutions.

We will hereafter assume that $N > 0$ and N is not a square. With these assumptions, it is always possible to show that $x^2 - Ny^2 = 1$ always has a solution other than $x = \pm 1$, $y = 0$. We will accept this result on faith and not prove it. There are two methods for proving the existence of a nontrivial solution: one depends on developing the extensive machinery of continued fractions, and the other (first constructed in 1842 by Dirichlet, who improved a proof given by Lagrange in 1766) is not short. Expositions of Dirichlet's proof can be found in the books by Griffin, LeVeque, Nagell, Shanks, and Uspensky and Heaslet mentioned in the Bibliography.

Because

$$x^2 - Ny^2 = (x + y\sqrt{N})(x - y\sqrt{N}),$$

irrational numbers of the form $x + y\sqrt{N}$ are closely connected with solutions of Fermat's equation. They also have several important properties, which we develop in the following lemmas. We will say that the irrational number

$$\alpha = r + s\sqrt{N}$$

(r and s are integers) *gives a solution* of $x^2 - Ny^2 = 1$ if and only if $r^2 - Ns^2 = 1$. For example, $3 + 2\sqrt{2}$ gives a solution of $x^2 - 2y^2 = 1$ and $8 + 3\sqrt{7}$ gives a solution of $x^2 - 7y^2 = 1$.

LEMMA 1. *If $N > 0$ is not a square, then*

$$x + y\sqrt{N} = r + s\sqrt{N}$$

if and only if $x = r$ and $y = s$.

PROOF. If $x = r$ and $y = s$, then clearly $x + y\sqrt{N} = r + s\sqrt{N}$. It is the converse that is important. To prove it, suppose that $x + y\sqrt{N} = r + s\sqrt{N}$ and $y \neq s$. Then

$$\sqrt{N} = \frac{x - r}{s - y}$$

is a rational number. But since N is not a square, $\sqrt{N}$ is irrational. It follows that $y = s$, and this implies $x = r$.

LEMMA 2. *For any integers a, b, c, d, N,*

$$(a^2 - Nb^2)(c^2 - Nd^2) = (ac + Nbd)^2 - N(ad + bc)^2.$$

PROOF. Multiply it out.

Exercise 3. Given that $2^2 - 3 \cdot 1^2 = 1$ and $7^2 - 3 \cdot 4^2 = 1$, use Lemma 2 to get another solution of $x^2 - 3y^2 = 1$.

LEMMA 3. *If α gives a solution of $x^2 - Ny^2 = 1$, then so does $1/\alpha$.*

PROOF. Let $\alpha = r + s\sqrt{N}$. Then we know that $r^2 - Ns^2 = 1$, and we have

$$\frac{1}{\alpha} = \frac{1}{r + s\sqrt{N}} \frac{r - s\sqrt{N}}{r - s\sqrt{N}} = \frac{r - s\sqrt{N}}{r^2 - Ns^2} = r - s\sqrt{N};$$

since $r^2 + N(-s)^2 = 1$, the lemma is proved.

LEMMA 4. *If α and β give solutions of $x^2 - Ny^2 = 1$, then so does $\alpha\beta$.*

PROOF. Let $\alpha = a + b\sqrt{N}$ and $\beta = c + d\sqrt{N}$. Then

$$\alpha\beta = (a + b\sqrt{N})(c + d\sqrt{N}) = (ac + Nbd) + (ad + bc)\sqrt{N}$$

and from Lemma 2 we have

$$(ac + Nbd)^2 - N(ad + bc)^2 = (a^2 - Nb^2)(c^2 - Nd^2) = 1,$$

and this shows that $\alpha\beta$ gives a solution.

LEMMA 5. *If α gives a solution of $x^2 - Ny^2 = 1$, then so does α^k for any integer k, positive, negative, or zero.*

Exercise 4. Prove Lemma 5. First show that it is true for all $k \geq 1$ by applying Lemma 4 and induction. Then show that it is true for $k \leq -1$ by applying Lemma 3. Then consider the case $k = 0$.

Lemma 5 shows that if we know one number α, $\alpha > 1$, which gives a solution of $x^2 - Ny^2 = 1$, then we can find infinitely many, namely those given by α^k, $k = 2, 3, \ldots$. The solutions are all different, because $\alpha^{k+1} > \alpha^k$ for all k. For example, $3 + 2\sqrt{2}$ gives a solution of $x^2 - 2y^2 = 1$. So, then, do

$$(3 + 2\sqrt{2})^2 = 17 + 12\sqrt{2}$$

and

$$(3 + 2\sqrt{2})^3 = (17 + 12\sqrt{2})(3 + 2\sqrt{2}) = 99 + 70\sqrt{2}$$

and higher powers of $3 + 2\sqrt{2}$.

Exercise 5. Check that $(3 + 2\sqrt{2})^2$ and $(3 + 2\sqrt{2})^3$ do give solutions of $x^2 - 2y^2 = 1$.

Exercise 6. $\alpha = 2 + \sqrt{3}$ gives a solution of $x^2 - 3y^2 = 1$. Find two other nontrivial solutions.

LEMMA 6. *Suppose that a, b, c, d are nonnegative and that $\alpha = a + b\sqrt{N}$ and $\beta = c + d\sqrt{N}$ give solutions of $x^2 - Ny^2 = 1$. Then $\alpha < \beta$ if and only if $a < c$.*

PROOF. Suppose that $a < c$. Then $a^2 < c^2$ and because $a^2 = 1 + Nb^2$ and $c^2 = 1 + Nd^2$ we have $Nb^2 < Nd^2$. Because none of b, d, N are negative, it follows that $b < d$. Together with $a < c$, this gives $\alpha < \beta$. Suppose that $\alpha < \beta$. If $a \geq c$, then $a^2 \geq c^2$. From this follows $b^2 \geq d^2$, which implies $\alpha \geq \beta$. Since this is impossible, we have $a < c$.

Now we are in a position to describe *all* the solutions of $x^2 - Ny^2 = 1$. Consider the set of all real numbers that give a solution of $x^2 - Ny^2 = 1$. Let θ be the smallest number in the set greater than one. Note that Lemma 6 guarantees that there will be such a smallest element, because the members $r + s\sqrt{N}$ of the set can be ordered according to the size of r, which is an integer, and any nonempty set of positive integers contains a smallest element. We will call θ the *generator* for $x^2 - Ny^2 = 1$, and we can now prove

THEOREM 1. *If θ is the generator for $x^2 - Ny^2 = 1$, then all nontrivial solutions of the equation with x and y positive are given by θ^k, $k = 1, 2, \ldots$.*

Note that the restriction of x and y to positive values loses us nothing essential, because nontrivial solutions come in quadruples

$$\{(x, y), (x, -y), (-x, y), (-x, -y)\},$$

and exactly one solution has two positive elements. Note also that we say nothing about the existence of a generator. It is a fact, as was noted earlier, that such a number can always be found. There is a method for getting θ from the continued fraction expansion of $\sqrt{N}$ by an easy calculation—easy in the sense that a computer would make light work of it. For some values of N, the computation is quite tedious. Of course, a generator can be found by trial, and for a long time, this was the only method available. In the seventh century, the Indian mathematician Brahmagupta said that a person who can within a year solve $x^2 - 92y^2 = 1$ is a true mathematician. Perhaps, and perhaps not; but such a person would at least be a true arithmetician, because the generator is $1151 + 120\sqrt{92}$. Solution by trial can also be difficult for so innocent-seeming an equation as $x^2 - 29y^2 = 1$; its smallest positive nontrivial solution is $x = 9801$ and $y = 1820$. The equation $x^2 - 61y^2$ has no positive nontrivial solution until $x = 1766319049$, $y = 226153980$. You can verify that this is a solution by multiplication, if you wish.

PROOF OF THEOREM 1. Let $x = r$, $y = s$ be any nontrivial solution of $x^2 - Ny^2 = 1$ with $r > 0$ and $s > 0$. Let $\alpha = r + s\sqrt{N}$. We want to show that $\alpha = \theta^k$ for some k. We know that $\alpha \geq \theta$ by the definition of generator, so there is a positive integer k such that

$$\theta^k \leq \alpha < \theta^{k+1}.$$

Thus $1 \leq \theta^{-k}\alpha < \theta$. From Lemmas 4 and 5, we know that $\theta^{-k}\alpha$ gives a solution of $x^2 - Ny^2 = 1$. We have defined θ to be the *smallest* number that is greater than one and which gives a nontrivial solution. But $\theta^{-k}\alpha$ is smaller than θ and also gives a solution. Hence $\theta^{-k}\alpha$ gives a trivial solution. Thus $\theta^{-k}\alpha = 1$ or $\alpha = \theta^k$, as we wanted to show.

Problems

1. Find the generator for $x^2 - Ny^2 = 1$ when N is (a) 6, (b) 7, (c) 8, (d) 10, (e) 11, (f) 12.

2. Find two positive nontrivial solutions of $x^2 - Ny^2 = 1$ when N is (a) 6, (b) 8, (c) 12, (d) 14, (e) 63, (f) 99.

3. Find three nontrivial solutions of $x^2 + 2xy - 2y^2 = 1$.

4. Determine infinitely many solutions of the equation in Problem 3.

5. (a) Show that if $a^2 > b$, then $x^2 + 2axy + by^2 = 1$ has infinitely many solutions if $a^2 - b$ is not a square.
   (b) If $a^2 < b$, show that the equation has only solutions in which $y = 0, 1,$ or -1.
   (c) What happens if $a^2 = b$?

6. (a) Let $a = 2mn$, $b = m^2 - n^2$, and $c = m^2 + n^2$ be the sides of a Pythagorean triangle. Suppose that $b = a + 1$. Show that $(m - n)^2 - 2n^2 = 1$, and determine all such triangles.
   (b) Find the smallest two such triangles.

7. (a) Show that a triangle with sides $2a - 1$, $2a$, $2a + 1$ has an integer area if and only if $3(a^2 - 1)$ is a square.
   (b) Find three such triangles.
   (c) Show that a triangle with sides $2a$, $2a + 1$, $2a + 2$ has a rational area if and only if $3((2a + 1)^2 - 4)$ is a square.
   (d) Show that this is impossible.

8. Show that if $x_1 + y_1\sqrt{N}$ is the generator for $x^2 - Ny^2 = 1$, then all solutions x_k, y_k can be written in the form

$$2x_k = (x_1 + y_1\sqrt{N})^k + (x_1 - y_1\sqrt{N})^k,$$
$$2\sqrt{N}y_k = (x_1 + y_1\sqrt{N})^k - (x_1 - y_1\sqrt{N})^k.$$

9. Show that if $x_1 + y_1\sqrt{N}$ is the generator of $x^2 - Ny^2 = 1$, then

$$0 < x_1 - y_1\sqrt{N} < 1.$$

10. With reference to the notation used in Problems 8 and 9, what happens to x_k/y_k as k gets larger and larger?

11. If $x_k + y_k\sqrt{2} = (3 + 2\sqrt{2})^k$, calculate $(x_k/y_k) - \sqrt{2}$ for $k = 1, 2, 3, 4$. ($\sqrt{2} = 1.414213562\ldots$.)

12. (a) Show that $x^2 - Ny^2 = -1$ has no solutions if $N \equiv 3 \pmod 4$.
    (b) Show that if x_1, y_1 is a solution of $x^2 - Ny^2 = -1$ with $x_1 > 1$, then u_k, u_k, $k = 1, 2, \ldots$ are solutions of $x^2 - Ny^2 = 1$, where

$$u_k + u_k\sqrt{N} = (x_1 + y_1\sqrt{N})^{2k}.$$

13. Show that if $x^2 - Ny^2 = k$ has one solution, then it has infinitely many.

14. $10^2 + 11^2 + 12^2 = 13^2 + 14^2$. Find another sum of three consecutive squares equal to a sum of two consecutive squares.

15. Show that $1 + n + n^2$ is never a square for $n > 0$.

16. If $x_1 + y_1\sqrt{N}$ is the generator for $x^2 - Ny^2 = 1$ and

$$x_k + y_k\sqrt{N} = (x_1 + y_1\sqrt{N})^k,$$

$k = 1, 2, \ldots$, show that

$$x_{k+1} = 2x_1x_k - x_{k-1},$$
$$y_{k+1} = 2x_1y_k - y_{k-1}.$$

17. Apply Problem 16 to extend the sequence of rational approximations to $\sqrt{2}$ found in Problem 11 to one more term.

Formulas for
Primes

In the earlier days of mathematics, there was a feeling that "function" and "formula" were more or less synonymous. Today, the notion of function is more general, but many of us still feel more comfortable with a function if we have an explicit formula to look at. There is no difference, really, between

$$\text{let } f(n) \text{ denote the largest prime factor of } n$$

and

$$\text{let } f(n) = \lim_{r \to 0} \lim_{s \to \infty} \lim_{t \to 0} \sum_{u=0}^{s} (1 - \cos^2 (u!)^r \pi/n)^{2t}.$$

The first expression is simpler, but perhaps the second lets us feel that we somehow have more control over f. (Some primitive people believe that if you know a man's name, then you have power over him. It is the same principle.)

The importance of formulas is of course not psychological but practical: in general, a formula will let us compute things of interest. Thus the formula above is *less* useful than the verbal description: it obscures what f is, and it does not lend itself to computation. But if we agree that formulas in general are nice things and worth having, then it is reasonable to search for them. We might ask for a formula for p_n, the nth prime. But the primes are so irregularly scattered through the integers that this is probably beyond all reason. The next best thing would be to have a formula that would produce nothing but primes. The aim of this section is to show that no very simple formula—no polynomial formula, in particular—will work, and then to exhibit a formula that produces primes for all n, $n = 1, 2, 3, \ldots$, but which is not, alas, adapted to computation:

$$f(n) = [\theta^{3^n}].$$

The simplest sort of formula to consider is

$$f(n) = an + b.$$

If we found such a function that gave nothing but primes, then we would have an arithmetic progression, with difference a, consisting entirely of primes. Looking through tables of primes, we can find various arithmetic progressions of primes, but none of infinite length: for example,

3, 5, 7;

7, 37, 67, 97, 127, 157;

199, 409, 619, 829, 1039, 1249, 1459, 1669, 1879, 2089.

Exercise 1. If $an + b$ is prime for two different values of n, show that a and b are relatively prime.

Exercise 2 (optional). If $2 \nmid a$, show that $an + b$ can be prime for no more than two consecutive values of n.

In spite of the above partial successes, no arithmetic progression can yield nothing but primes, as we now show. Suppose that $an + b = p$, where p is a prime. Let $n_k = n + kp$, $k = 0, 1, \ldots$. Then the n_kth term of the progression is

$$an_k + b = a(n + kp) + b = (an + b) + akp = p + akp,$$

and this is divisible by p for all k. Thus every pth term of the progression is divisible by p (because the numbers n_k come at intervals of p), and thus the progression contains infinitely many composite numbers.

A sequence $\{an + b\}$ cannot consist entirely of primes, but it is natural to ask whether the sequence can contain infinitely many primes. The answer to this is given by Dirichlet's Theorem: If $(a, b) = 1$, then the sequence $\{an + b\}$ contains infinitely many primes. For example, among the members of the sequence $\{4n + 1\}$ are the primes 5, 13, 17, 29, 37, 41, $\ldots$, and among $\{12n + 7\}$ are 7, 19, 31, 43, 67, $\ldots$; Dirichlet's Theorem says that we will never come to a last prime in either sequence. The condition $(a, b) = 1$ is clearly necessary: $\{6n + 3\}$ contains only one prime and $\{6n + 4\}$ contains none. Dirichlet's great achievement was in showing that the condition was also sufficient. The proof of this theorem is not at all easy, and we will not attempt it. Even so, it is possible to prove not only that there are infinitely many primes in the sequence, but to estimate how they are distributed. The number of primes in the sequence $\{an + b\}$, where $(a, b) = 1$, which are less than N is roughly equal to $N/\phi(a) \log N$, where ϕ denotes the Euler ϕ-function.

We now prove a theorem which shows that arithmetic progressions act in a way like the sequence of integers; it also lets us see how many consecutive prime terms a progression can contain.

THEOREM 1. *If* $p \nmid a$, *then every pth term of the sequence* $\{an + b\}$ *is divisible by* p.

PROOF. Since $p \nmid a$, p and a are relatively prime, and so there exist integers r and s such that $pr + as = 1$. Let

$$n_k = kp - bs, \qquad k = 1, 2, \ldots .$$

Then

$$an_k + b = a(kp - bs) + b = akp - bas + b$$
$$= akp - b(1 - pr) + b = akp - b + bpr + b$$
$$= p(ak + br).$$

Thus $p \mid (an_k + b)$ for every k, $k = 1, 2, \ldots .$ Since $n_{k+1} - n_k = p$, the terms $an_k + b$ occur p terms apart.

From Theorem 1, it follows that if $2 \nmid a$, then every other term in the sequence $\{an + b\}$ is divisible by two. Hence, the sequence cannot have more than three consecutive non-composite numbers, and this can happen only if the second of the three is 2. In general, if $p \nmid a$, then $\{an + b\}$ cannot have any more than $2p - 1$ consecutive terms that are not composite, and this can happen only if the middle term of the $2p - 1$ is p itself. In all other cases, the upper limit is $p - 1$. Thus, if we want to search for an arithmetic progression that contains 12 consecutive primes, the common difference a cannot be picked at random. If 2, 3, 5, 7, and 11 do not appear in the progression, the common difference must be divisible by $2 \cdot 3 \cdot 5 \cdot 7 \cdot 11 = 2310$. The largest arithmetic progression now known that consists entirely of primes has 16 terms: $2236133941 + 223092870n$, $n = 0, 1, \ldots, 15$. See "12 to 16 Primes in Arithmetic Progression," by Edgar Karst [*J. of Recreational Mathematics* 2 (1969), 214–215].

Since linear formulas have failed as prime-producing functions, the next thing to try would be quadratics: can $an^2 + bn + c$ be prime for all integers n? Again, we can get partial successes. For example, if $f(n) = n^2 + 21n + 1$, then $f(n)$ is not composite for $n = -38, -37, \ldots, 17$; a string of 56 consecutive integers. But $f(18) = 703 = 37 \cdot 19$. (Note that "not composite" and "prime" are not synonymous, for 1 is neither composite nor prime. Since $f(0) = 1$, we cannot say that $f(n)$ is prime for 56 consecutive integers.)

Exercise 3. Use congruences to show that $2 \nmid f(n)$, $3 \nmid f(n)$, and $7 \nmid f(n)$ for any n.

Exercise 4. Show that $19 \mid f(n)$ if $n \equiv -1 \pmod{19}$.

Exercise 5 (optional). Show that $19 \mid f(n)$ only if $n \equiv -1 \pmod{19}$.

Another well-known example of a prime-rich quadratic is $n^2 + n + 41$, which is not only not composite, but prime for 80 consecutive integers: $n = -40, -39, \ldots, 39$. This is the present champion: no quadratic is known that will produce more than 80 noncomposite numbers in a row. Since the primes thin out as numbers increase, it is not unreasonable to conjecture that no quadratic can do better, and that 80 is the absolute maximum number of consecutive noncomposites that a quadratic can produce. There is little likelihood that this conjecture will be settled any time soon, if ever. All that is known is that no quadratic of the form $n^2 + n + A, A > 41$, gives primes for $n = 0, 1, \ldots, A - 2$, and this is a consequence of a difficult result proved only in 1967.

No quadratic can always be a prime. Suppose that

$$f(n) = an^2 + bn + c = p$$

is prime for some n. That is, $an^2 + bn + c \equiv 0 \pmod{p}$. Let $n_k = n + kp$, $k = 0, 1, \ldots$. Then

$$f(n_k) = a(n + kp)^2 + b(n + kp) + c$$
$$= an^2 + 2ankp + ak^2p^2 + bn + bkp + c$$

or

$$f(n_k) \equiv an^2 + bn + c \equiv 0 \pmod{p},$$

so $f(n_k)$ is divisible by p for every k. Thus every pth term of the sequence $\{an^2 + bn + c\}$ is divisible by p, so the sequence contains infinitely many composite numbers.

After seeing Dirichlet's Theorem, we might wonder if $\{an^2 + bn + c\}$ contains infinitely many primes if a, b, and c have no common factor. No theorem corresponding to Dirichlet's Theorem is known for quadratics. In fact, it has not yet been proved that $n^2 + 1$ is prime infinitely often, though it seems unlikely that this should not be so.

We might suspect that cubic polynomials are no better than quadratics for representing primes, and we would be right. If f is a polynomial of any degree and $f(n) = p$, p a prime, for some n, then it can be shown that $p \mid f(n + kp)$ for all k, $k = 0, 1, \ldots$, in exactly the same way as it was shown for linear and quadratic polynomials. Further, it is known that if f and g are polynomials, then it is impossible for $[f(n)/g(n)]$ to be prime for all n, unless it becomes constant for n sufficiently large.

On the other hand, we can construct a polynomial that assumes as many consecutive prime values as we want, because it can be shown that it is always possible to make a polynomial of degree d take on $d + 1$ arbitrarily assigned values. For example, if

$$60f(x) = 7x^5 - 85x^4 + 355x^3 - 575x^2 + 418x + 180,$$

then we have

n	0	1	2	3	4	5
$f(n)$	3	5	7	11	13	17.

A similar polynomial could be constructed to take on 81 consecutive prime values, but it would be of degree 80.

After giving up on polynomials, it would be natural to try exponential functions. For example, if

$$f(n) = [(3/2)^n],$$

then $f(n)$ is prime for $n = 2, 3, 4, 5, 6, 7$ (the values of the function are 2, 3, 5, 7, 11, and 17), but $f(8) = 25$, and the next prime in the sequence does not come until $f(21) = 4987$. No one has proved that a formula like $f(n) = [\theta^n]$ cannot always give a prime. Nor is it known whether $[\theta^n]$ can be prime infinitely often. Such questions seem hopelessly difficult.

Nevertheless, there do exist functions, expressible as a simple formula, that always represent primes. We will prove, partly, a striking result of W. H. Mills:

THEOREM 2. *There is a real number θ such that $[\theta^{3^n}]$ is a prime for all n, $n = 1, 2, \ldots$.*

As we shall see, this theorem contains less than meets the eye, and it should not seem nearly so striking after we finish the proof. The proof gives a construction for θ, but the construction depends on being able to recognize arbitrarily large primes. If we could recognize arbitrarily large primes, we would have no need of the formula.

In the proof we will use two theorems from analysis.

THEOREM. *If a sequence u_1, u_2, u_3, $\ldots$, u_n, $\ldots$ is bounded above and nondecreasing, then it has a limit, θ, as n increases without bound.*

That is, if there is a number M such that $u_n < M$ for all n and $u_n \leq u_{n+1}$ for all n, $n = 1, 2, \ldots$, then there is a number θ such that the difference between θ and u_n becomes arbitrarily small as n increases without bound. We will not prove this theorem, or the next.

THEOREM. *If a sequence $v_1, v_2, v_3, \ldots, v_n, \ldots$ is bounded below and non-increasing, then it has a limit, ϕ, as n increases without bound.*

We will write

$$\lim_{n \to \infty} u_n = \theta \quad \text{and} \quad \lim_{n \to \infty} v_n = \phi$$

and read "the limit of u_n as n approaches infinity equals θ," and a corresponding statement for v_n.

PROOF OF THEOREM 2. The proof depends on the following theorem: there is an integer A such that if $n > A$, then there is a prime p such that

(1) $$n^3 < p < (n+1)^3 - 1.$$

We will not prove this—its proof depends on deep properties of the Riemann ζ-function. We will use (1) to determine a sequence of primes that will in turn determine θ. Let p_1 be any prime greater than A, and for $n = 1, 2, \ldots$, let p_{n+1} be a prime such that

(2) $$p_n^3 < p_{n+1} < (p_n + 1)^3 - 1.$$

Such a prime exists for each n on account of (1). Let

(3) $$u_n = p_n^{3^{-n}} \quad \text{and} \quad v_n = (p_n + 1)^{3^{-n}},$$

$n = 1, 2, \ldots$. We see that as n increases, u_n increases, because from (2),

(4) $$u_{n+1} = p_{n+1}^{3^{-n-1}} > (p_n^3)^{3^{-n-1}} = p_n^{3^{-n}} = u_n.$$

Furthermore, $\{v_n\}$ is a decreasing sequence, because from (2),

(5) $$v_{n+1} = (p_{n+1} + 1)^{3^{-n-1}} < ((p_n + 1)^3 - 1 + 1)^{3^{-n-1}} = (p_n + 1)^{3^{-n}} = v_n.$$

It is clear from (3) that $u_n < v_n$. Hence, because of (5),

$$u_n < v_n < v_{n-1} < \cdots < v_1,$$

so $u_n < v_1$ for all n. In the same way, from (4) we have

$$v_n > u_n > u_{n-1} > \cdots > u_1,$$

so $v_n > u_1$ for all n. Thus $\{u_n\}$ is an increasing sequence of numbers that is bounded above by v_1. It follows that $\{u_n\}$ has a limit. Call it θ. Also, $\{v_n\}$ is a decreasing sequence of numbers that is bounded below by u_1. Hence

$\{v_n\}$ has a limit too. Call it ϕ. Since $u_n < v_n$ for all n, it follows that $\theta \leq \phi$. In fact, since $\{u_n\}$ increases and $\{v_n\}$ decreases, we have

$$u_n < \theta \leq \phi < v_n$$

for all n; thus

$$u_n^{3^n} < \theta^{3^n} \leq \phi^{3^n} < v_n^{3^n}$$

for all n. But from the definitions of u_n and v_n,

$$u_n^{3^n} = p_n \quad \text{and} \quad v_n^{3^n} = p_n + 1.$$

Thus

$$p_n < \theta^{3^n} < p_n + 1.$$

This locates θ^{3^n} between two consecutive integers, and so

$$[\theta^{3^n}] = p_n,$$

a prime, for all n.

From the construction, we see that knowledge of θ and knowledge of all the primes is essentially equivalent, so that the theorem gives us nothing that we did not have before, except perhaps pleasure at seeing a clever idea neatly worked out. The theorem would be important only if we could discover what θ is by some method independent of all the primes, and this is not likely.

Problems

1. Find a quadratic polynomial f such that $f(0) = 2$, $f(1) = 3$, and $f(2) = 5$.

2. Write a formula $y = f(n)$ such that y is a prime for *all n, $n = 1, 2, \ldots$*.

3. Let $f(n) = n^2 + 21n + 1$. Show that $7 \nmid f(n)$ and $11 \nmid f(n)$ for any n.

4. Show that $n^3 + 11n + 1$ is never divisible by 2, 3, 5, or 7.

5. Which primes can divide $n^2 + 2$?

6. Which primes can divide $n^2 + 2n + 3$?

7. What values can a have if $0 < a \leq 100$ and $an^2 + n + 1$ is never divisible by 2, 3, or 5?

8. Show that if an odd prime p divides $n^2 + n + 41$ for some n, then

$$(p + 1)^2/4 - 41$$

is a quadratic residue modulo p.

9. Show that if p is a prime and $p \mid a^n$ for some a and $n > 0$, then $p \mid (a + kp)^n$ for $k = 0, 1, \ldots$.

10. If f is a polynomial and $f(a) = p$, a prime, then show that $p \mid f(a + kp)$ for all k.

11. Let $f(n) = \sin \pi((1 + (n - 1)!)/n)$. Show that $f(n) = 0$ if and only if n is a prime.

12. Let p_n denote the nth prime, and let

$$\theta = \sum_{n=1}^{\infty} p_n/10^{n^2} = .2003000050000007\ldots.$$

Show that

$$p_n = [10^{n^2}\theta] - 10^{2n-1}[10^{(n-1)^2}\theta].$$

13. Show that the sum of the first n positive integers divides their product if and only if $n + 1$ is not an odd prime.

14. Show that an odd prime can be written as a sum of more than one consecutive positive integers in exactly one way.

15. Consider the infinite array whose first five lines are

$$
\begin{array}{ccccccccccccccccc}
 & & & & & & & 1 & & 1 & & & & & & & \\
 & & & & & & 1 & & 2 & & 1 & & & & & & \\
 & & & & & 1 & & 3 & & 2 & & 3 & & 1 & & & \\
 & & & & 1 & & 4 & & 3 & & 5 & & 2 & & 5 & & 3 & & 4 & & 1 & & \\
 & 1 & & 5 & & 4 & & 7 & & 3 & & 8 & & 5 & & 7 & & 2 & & 7 & & 5 & & 8 & & 3 & & 7 & & 4 & & 5 & & 1 \\
\end{array}
$$

(a) How is it formed?

(b) Show that the number of terms in the nth line is $2^{n-1} + 1$ and their sum is $3^{n-1} + 1$.

(c) Prove that any two consecutive entries in any line are relatively prime.

(d) Show that n is prime if and only if it occurs $n - 1$ times in the nth line.

16. Catalan observed (1876) that if $p_0 = 2$ and

$$p_{n+1} = 2^{p_n} - 1,$$

then p_k is a prime for $k = 1, 2, 3, 4$. Perhaps it is prime for all k. That is a difficult conjecture, for $\{p_k\}$ increases very rapidly. About how many digits has p_5?

Bounds for π (x)

Although we may have a hard time telling whether a specific large number is a prime, we can make quite accurate statements about *how many* numbers in a given interval are prime. It is the same idea, almost, as the actuary's mortality table, which can predict very closely the number of people aged 72 who will die next year, but which is no help in singling out the doomed lives. Let $\pi(x)$ denote the number of primes less than or equal to x. (*Note:* We will not restrict x, y, and z to integer values in this section. Other lower-case italic letters—except $e = 2.7182818284590452...$—will continue to denote integers.) It is the purpose of this section to derive bounds for $\pi(x)$ when x is large. In doing so, we will have to use some properties of the natural logarithm function. If we denote by e the limit of the sequence $\left\{\left(1 + \dfrac{1}{n}\right)^n\right\}$, $n = 1, 2, \ldots$, as n increases without bound, then the *natural logarithm* of x (denoted by $\log x$ or $\ln x$) can be defined for any $x > 0$ by

$$x = e^{\log x}.$$

It has the properties that for any positive numbers x and y and any z,

$$\log xy = \log x + \log y$$
$$\log x^z = z \log x.$$

Moreover,

$$\lim_{x \to \infty} (\log x)/x = 0.$$

The natural logarithm of x increases as x increases, but at a decreasing rate.

A portion of its graph is given in the figure below and some numerical values are, approximately,

x	.01	.1	1	2	3	10	500	1,000,000
$\log x$	-4.6	-2.3	0	.7	1.1	2.3	6.2	13.8

In addition to these properties of the natural logarithm, we will use one result (Lemma 6) without proof. Using these properties and Lemma 6, we will show that $\pi(x)$ increases at roughly the same rate as $x/\log x$

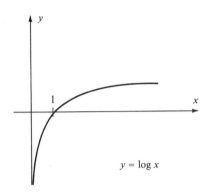

$y = \log x$

Exercise 1. What is $\pi(8)$? $\pi(12)$? $\pi(3.2)$? $\pi(\pi)$?

That $\pi(x)$ increases at roughly the same rate as $x/\log x$ was induced from the observation of numerical evidence. If you take a table of $\pi(x)$, such as this

x	100	200	300	400	500	600	700	800	900	1000
$\pi(x)$	25	46	62	78	95	109	125	139	154	168

x	10,000	100,000	1,000,000	10,000,000	100,000,000
$\pi(x)$	1,229	9,592	78,498	664,579	5,761,455,

and calculate the ratio of $\pi(x)$ to $x/\log x$, you will find that it stays between .9 and 1.2 and seems to approach 1 as x increases. What we will prove in this section is the comparatively weak result

(1) $$\frac{1}{3} < \frac{\pi(x) \log x}{x} < \frac{10}{3} \qquad \text{when} \qquad x > 400{,}000.$$

Before we start, we will mention what is known to be true. In fact,

(2) $$\lim_{x \to \infty} \frac{\pi(x) \log x}{x} = 1;$$

this is the famous Prime Number Theorem. The inequalities in (1) were proved (with rather better constants than 400,000, 1/3, and 10/3) by Tchebyshev in 1850. The history of (2) starts in 1780, when Legendre guessed that

$$f(x) = \frac{x}{\log x - 1.08366}$$

was a good approximation to $\pi(x)$—probably in the sense that, as in (2),

$$\lim_{x \to \infty} \pi(x)/f(x) = 1.$$

In 1792, Gauss suggested a function that fits the facts much better than Legendre's guess. In 1859, Riemann made an attempt to prove (2). His proof was not complete, but it contained the ideas necessary for a complete proof. Even so, it was not until 1896 that Hadamard and De la Vallée Poussin independently proved the Prime Number Theorem. Work on refinements of it still continues today.

We will start on the road to the proof of (1) by introducing two new functions. Let

$$\theta(x) = \sum_{p \le x} \log p.$$

This may seem to be a less natural function than

$$\pi(x) = \sum_{p \le x} 1,$$

but it turns out to be easier to work with. A relation between the two is given in

LEMMA 1. *For all $x > 1$,*

$$\frac{\theta(x) - \theta(x^{1/2})}{\log x} \le \pi(x) - \pi(x^{1/2}) \le \frac{2(\theta(x) - \theta(x^{1/2}))}{\log x}.$$

PROOF. We have

$$\sum_{x^{1/2} < p \le x} \log x^{1/2} \le \sum_{x^{1/2} < p \le x} \log p \le \sum_{x^{1/2} < p \le x} \log x$$

or

(3) $$\tfrac{1}{2} \log x \sum_{x^{1/2} < p \le x} 1 \le \sum_{x^{1/2} < p \le x} \log p \le \log x \sum_{x^{1/2} < p \le x} 1.$$

But

$$\sum_{x^{1/2} < p \le x} 1 = \sum_{p \le x} 1 - \sum_{p \le x^{1/2}} 1 = \pi(x) - \pi(x^{1/2})$$

and

$$\sum_{x^{1/2} < p \le x} \log p = \theta(x) - \theta(x^{1/2}).$$

Thus (3) becomes

$$\tfrac{1}{2}\left(\pi(x) - \pi(x^{1/2})\right)\log x \le \theta(x) - \theta(x^{1/2}) \le \left(\pi(x) - \pi(x^{1/2})\right)\log x,$$

which is equivalent to what we wanted to show.

Later we will see that $\pi(x^{1/2})$ is negligible in comparison with $\pi(x)$ (we mean that

$$\lim_{x \to \infty} \frac{\pi(x^{1/2})}{\pi(x)} = 0),$$

and the same holds for $\theta(x^{1/2})$ in comparison with $\theta(x)$. If we neglect these terms, then the lemma says that $\pi(x)$ and $\theta(x)/\log x$ increase at the same rate. Thus, to show that $\pi(x)$ increases at the same rate as $x/\log x$, we need only show that $\theta(x)$ increases at the same rate as x. We will find it easier to deal with a function that looks even worse than θ: let

(4) $$\psi(x) = \theta(x) + \theta(x^{1/2}) + \theta(x^{1/3}) + \cdots.$$

Note that the sum is finite, since if $x^{1/m} < 2$, then

$$\theta(x^{1/m}) = \sum_{p \le x^{1/m}} \log p = 0,$$

because the sum has no terms in it.

Exercise 2. Show that the last nonzero term in (4) is $\theta(x^{1/m})$, where $m = [\log x/\log 2]$.

From Exercise 2 it follows that the sum in (4) contains no more than $2 \log x$ terms.

Exercise 3. Demonstrate that this is so: show that $[\log x/\log 2] \le 2 \log x$.

Exercise 4 (optional). Calculate $\psi(32)$.

We now prove a lemma that relates ψ with θ. This is in a sense the weakest of our lemmas, and as you will see later, it is the reason why x must be larger than 400,000 in Theorem 1.

LEMMA 2. *For all $x \ge 1$,*

$$\psi(x) - x^{1/2}\log^2 x \le \theta(x) \le \psi(x).$$

PROOF. First we get an upper bound for $\theta(x^{1/m})$. By its definition,

$$\theta(x^{1/m}) = \sum_{p \le x^{1/m}} \log p.$$

The largest summand is no greater than $\log x$, and there are at most $\pi(x^{1/m})$ terms in the sum. Since $\pi(x^{1/m}) \leq x^{1/m}$ (because there are at least as many integers as primes), we have

$$\theta(x^{1/m}) \leq x^{1/m} \log x^{1/m}.$$

Putting this in (4), we get

$$\psi(x) - \theta(x) = \theta(x^{1/2}) + \theta(x^{1/3}) + \cdots \leq x^{1/2} \log x^{1/2} + x^{1/3} \log x^{1/3} + \cdots.$$

Each term on the right-hand side of the last inequality is bounded above by $x^{1/2} \log x^{1/2}$ and, as we have already seen, there are no more than $2 \log x$ terms. It follows that

$$\psi(x) - \theta(x) \leq (x^{1/2} \log x^{1/2})2 \log x = x^{1/2} \log^2 x,$$

and so

$$\theta(x) \geq \psi(x) - x^{1/2} \log^2 x.$$

From the definition of ψ we see that for all x,

$$\theta(x) \leq \psi(x),$$

and these last two inequalities prove the lemma.

Lemma 2 shows that if $\psi(x)$ increases at the same rate as x, then so does $\theta(x)$. What we are aiming at is this: using $\psi(x)$, we will construct a function that can be put into a form that does not depend on the primes. We can then get fairly precise estimates on how fast it grows. Then we relate its rate of growth to that of ψ, and use Lemmas 1 and 2 to relate it back to the growth of π. Let $S(x)$ be defined by

$$(5) \qquad S(x) = \psi(x) + \psi(x/2) + \psi(x/3) + \cdots.$$

This is the function that can be put into a form independent of the primes. Using (4), we can write $S(x)$ in the form

$$S(x) = \theta(x) + \theta(x^{1/2}) + \theta(x^{1/3}) + \cdots$$
$$(6) \qquad\qquad + \theta(x/2) + \theta((x/2)^{1/2}) + \theta((x/2)^{1/3}) + \cdots$$
$$\qquad\qquad + \theta(x/3) + \theta((x/3)^{1/2}) + \theta((x/3)^{1/3}) + \cdots.$$

Note that all the sums in (5) and (6) are finite. In each row, there will be a term beyond which the argument of the function is so small that the function is zero from that term on.

Exercise 5 (optional). Approximately how many nonzero terms are there on the right-hand side of (6)?

"Why," you may be thinking (and ought to be), "would anyone ever think to consider such a function as $S(x)$?" Tchebyshev did, and if you had the talent of Tchebyshev, and worked on the problem of estimating $\psi(x)$ with sufficient diligence and inspiration for sufficiently long, you would most likely be led to consider this sum too. It is really quite natural. Hindsight shows this conclusively.

Exercise 6. What is the last nonzero entry in the second row of (6)? What is the last nonzero entry in the second column?

We will now count how many times $\log p$ appears on the right-hand side of (6). First, let us see how many times it appears in the first column: $\log p$ appears in the sum for $\theta(x/m)$ if and only if $p \leq x/m$. That is, it appears in the sums for $\theta(x/m)$ for those values of m such that $m \leq x/p$. There are $[x/p]$ such integers. In the second column $\log p$ appears in the sum for $\theta((x/m)^{1/2})$ if and only if $p \leq (x/m)^{1/2}$, or, if and only if $m \leq x/p^2$. Thus it appears in the sum for $\theta((x/m)^{1/2})$ for

$$m = 1, 2, \ldots, [x/p^2].$$

In the same way, we can show that $\log p$ appears in the third column of (6) $[x/p^3]$ times. And so on: we have proved

LEMMA 3. *$\log p$ appears in the sum for $S(x)$ exactly $S_p(x)$ times, where*

$$S_p(x) = [x/p] + [x/p^2] + [x/p^3] + \cdots .$$

Exercise 7. Is this a finite sum too?

Exercise 8 (optional). Show that

$$\sum_{m \leq x} \theta(x/m) = \sum_{p \leq x} [x/p] \log p.$$

LEMMA 4. *The highest power of p that divides $n!$ is $S_p(n)$.*

PROOF. Each multiple of p less than or equal to n adds one power of p to $n!$, and there are $[n/p]$ such multiples. The multiples of p^2 each contribute an additional power of p, and there are $[n/p^2]$ such multiples. And so on: the additional contribution made by the multiples of p^k is $[n/p^k]$, and hence p to the power

$$[n/p] + [n/p^2] + [n/p^3] + \cdots$$

exactly divides $n!$.

Exercise 9 (optional). Exactly how many terms are there in the sum of Lemma 4?

Exercise 10. How many times does 11 go into 1111!?

Exercise 11. Show that $[x/n] = [[x]/n]$.

With the aid of Lemma 4, we can put $S(x)$ into a form that does not explicitly involve any primes:

LEMMA 5. *For all $x \geq 1$,*

$$S(x) = \sum_{n \leq x} \log n.$$

This can be regarded as the key step in the estimation of $\pi(x)$. After this lemma is proved, we have to find out the rate of growth of $S(x)$, but this is a more or less routine problem. After this has been done, we have to go back and relate $S(x)$ with $\pi(x)$, but this is also routine.

PROOF OF LEMMA 5. From (6), we know that $S(x)$ is the sum of a number of terms of the form $\log p$, and from Lemma 3 we know how many times $\log p$ appears in the sum: exactly $S_p(x)$ times. Thus

(7) $S(x) = S_2(x) \log 2 + S_3(x) \log 3 + S_5(x) \log 5 + \cdots$

$= \log(2^{S_2(x)} 3^{S_3(x)} 5^{S_5(x)} \cdots).$

On the other hand,

(8) $[x]! = 2^{e_2(x)} 3^{e_3(x)} 5^{e_5(x)} \cdots,$

where, according to Lemma 4,

$$e_p(x) = [[x]/p] + [[x]/p^2] + [[x]/p^3] + \cdots .$$

But from Exercise 11, $[[x]/p^k] = [x/p^k]$ for any p and k, so it follows that $e_p(x) = S_p(x)$. With this fact, we can compare (7) and (8) and see that

$$S(x) = \log[x]! = \sum_{1 \leq n \leq [x]} \log n = \sum_{n \leq x} \log n,$$

and the lemma is proved.

Exercise 12 (optional). Calculate, approximately, $S(x)$ for some values of x and guess its rate of growth.

In order to find out how fast $S(x)$ grows, we need a result from analysis, which we will accept without proof:

LEMMA 6. *For any integer n, $n \geq 7$,*

$$\left(\frac{n}{e}\right)^n \leq n! \leq n\left(\frac{n}{e}\right)^n.$$

This lemma gives a weak form of a well-known result (Stirling's formula), and it can be proved by induction if you recall the definition of e, the base of natural logarithms. From it follows

LEMMA 7. *For any integer n, $n \geq 7$,*

$$n \log n - n \leq S(n) \leq n \log n - n + \log n.$$

PROOF. Because

$$S(n) = \sum_{k \leq n} \log k = \log n!,$$

we have from Lemma 6,

$$S(n) \leq \log n \left(\frac{n}{e}\right)^n = \log n + n(\log n - \log e)$$

$$= n \log n - n + \log n$$

and

$$S(n) \geq \log\left(\frac{n}{e}\right)^n = n \log n - n.$$

Lemma 7 says that $S(x)$ behaves, roughly, like $x \log x$ when x is an integer. This remains true when x is not an integer. Routine calculations will prove

LEMMA 8. *For any x, $x > 7$,*

$$x \log x - x - \log x - 1 \leq S(x) \leq x \log x - x + \log x.$$

We omit the details of the proof, as they are neither interesting nor instructive. Lemma 8 shows that $S(x)$ increases at the same rate as $x \log x$. We are trying to show that $\psi(x)$ increases at the same rate as x. More computation is needed, but to start with, we need another burst of inspiration. We borrow it from Tchebyshev. Consider $S(x) - 2S(x/2)$. From the definition of $S(x)$ we have

$$(9) \qquad S(x) - 2S(x/2) = \psi(x) - \psi(x/2) + \psi(x/3) - \psi(x/4) + \cdots.$$

From Lemma 8, we can get upper and lower bounds for $S(x) - 2S(x/2)$: when $x > 7$,

$$S(x) - 2S(x/2) \leq x \log x - x + \log x - 2\left(\frac{x}{2}\log\frac{x}{2} - \frac{x}{2} - \log\frac{x}{2} - 1\right)$$

$$= x \log 2 + 3 \log x + 2 - 2 \log 2.$$

In the other direction, we have

$$S(x) - 2S(x/2) \geq x \log x - x - \log x - 1 - 2\left(\frac{x}{2}\log\frac{x}{2} - \frac{x}{2} + \log\frac{x}{2}\right)$$

$$= x \log 2 - 3 \log x + 2 \log 2 - 1.$$

We collect the last two inequalities in

LEMMA 9. *For $x > 7$,*

$$x \log 2 - 3 \log x + 2 \log 2 - 1 \leq S(x) - 2S(x/2)$$
$$\leq x \log 2 + 3 \log x + 2 - 2 \log 2.$$

We are almost done. Notice that as x decreases, $\psi(x)$ does not increase because

$$\psi(x) = \theta(x) + \theta(x^{1/2}) + \theta(x^{1/3}) + \cdots,$$

and as x decreases, $\theta(x^{1/m})$ does not increase, for each m. Thus

$$S(x) - 2S(x/2) = \psi(x) - \psi(x/2) + \psi(x/3) - \cdots$$

is an alternating series with nonincreasing terms. So,

$$S(x) - 2S(x/2) = \psi(x) - (\psi(x/2) - \psi(x/3)) - (\psi(x/4) - \psi(x/5)) - \cdots$$

shows that

$$S(x) - 2S(x/2) \leq \psi(x);$$

and

$$S(x) - 2S(x/2) = (\psi(x) - \psi(x/2)) + (\psi(x/3) - \psi(x/4)) + \cdots$$

shows that

$$S(x) - 2S(x/2) \geq \psi(x) - \psi(x/2).$$

Thus we have

(10) $$\psi(x) - \psi(x/2) \leq S(x) - 2S(x/2) \leq \psi(x).$$

The right-hand inequality in (10) gives a lower bound for $\psi(x)$: we have from Lemma 9 that

$$\psi(x) \geq S(x) - 2S(x/2) \geq x \log 2 - 3 \log x + 2 \log 2 - 1.$$

From this, we can conclude

LEMMA 10. *For $x > 150$, $\psi(x) \geq x/2$.*

PROOF. It is clear that $(3 \log x + 2 \log 2 - 1)/x$ decreases to zero as $x \to \infty$. If $x > 150$, then (because $e^5 = 148.4 \ldots < 150$),

$$\frac{3 \log x - 2 \log 2 + 1}{x} \leq \frac{3 \cdot 5 - 1 \cdot 4 + 1}{150} \leq .1.$$

Thus, when $x > 150$,

$$x \log 2 - 3 \log x + 2 \log 2 - 1 \geq x(\log 2 - .1) \geq x/2.$$

It is also clear that for x sufficiently large, we can prove that $\psi(x) \geq x(\log 2 - \epsilon)$ for any $\epsilon > 0$. It is not worth the trouble, because from the Prime Number Theorem it follows that $\psi(x)/x \to 1$ as $x \to \infty$, and thus $\psi(x) \geq x(1 - \epsilon)$ will be true for any $\epsilon > 0$, when x is sufficiently large.

To get an upper bound for ψ, we need one more clever trick. It is contained in the next two lemmas.

LEMMA 11. *For $x > 403$, $\psi(x) - \psi(x/2) \leq 3x/4$.*

PROOF. From (10) and Lemma 9 we have

$$\psi(x) - \psi(x/2) \leq S(x) - 2S(x/2) \leq x \log 2 + 3 \log x + 2 - 2 \log 2.$$

It is clear that $(3 \log x + 2 - 2 \log 2)/x$ decreases to 0 as $x \to \infty$. If $x > 403$ (this number was chosen because $e^6 = 403.4 \ldots$), then the fraction is at most

$$\frac{18 + 2 - 1.4}{403} < .05.$$

Thus, for $x > 403$, we have

$$x \log 2 + 3 \log x + 2 - 2 \log 2 < x(.694 + .05) < 3x/4,$$

which proves the lemma.

Lemma 11 allows us to prove

LEMMA 12. *If $x > 403$, then $\psi(x) \leq 3x/2$.*

PROOF. From Lemma 11 we have

$$\psi(x) - \frac{3x}{2} \leq \psi(x/2) - \frac{3x}{4}.$$

If we apply this over and over, we get

$$\psi(x) - \frac{3x}{2} \leq \psi(x/2) - \frac{3x}{4} \leq \psi(x/8) - \frac{3x}{16} \leq \cdots \leq \psi(x/2^n) - \frac{3x}{2^{n+1}},$$

for any integer n. If n is large enough, then $\psi(x/2^n) = 0$.

Exercise 13 (optional). How large is large enough?

With this n, we have

$$\psi(x) - \frac{3x}{2} \leq \frac{-3x}{2^{n+1}} \leq 0,$$

and this proves the lemma.

At last, we can prove

THEOREM 1. *If x > 400,000, then*

$$\frac{1}{3}\frac{x}{\log x} \leq \pi(x) \leq \frac{10}{3}\frac{x}{\log x}.$$

PROOF. We have

$$\theta(x) = \sum_{p \leq x} \log p \geq \sum_{x^{1/2} < p \leq x} \log p.$$

There are $\pi(x) - \pi(x^{1/2})$ terms in the last sum and the smallest term is not less than $\log x^{1/2}$. Thus

$$\theta(x) \geq (\pi(x) - \pi(x^{1/2}))\log x^{1/2} \geq (\pi(x) - x^{1/2})\log x^{1/2}$$

or

$$\pi(x) - x^{1/2} \leq \frac{2\theta(x)}{\log x}.$$

From the definition of ψ, we know that $\psi(x) \geq \theta(x)$, so

$$\pi(x) \leq \frac{2\psi(x)}{\log x} + x^{1/2}.$$

Applying Lemma 12, we have that for $x > 403$,

$$\pi(x) \leq \frac{2}{\log x} \cdot \frac{3x}{2} + x^{1/2}.$$

Furthermore, it is easy to see that when $x > 400,000$,

$$x^{1/2} < \frac{.01x}{\log x}.$$

Exercise 14 (optional). Verify this.

We thus have, when $x > 400,000$,

$$\pi(x) \leq \frac{3x}{\log x} + \frac{.01x}{\log x} < \frac{10}{3}\frac{x}{\log x}.$$

This proves half of the theorem. To prove the other half, we have

$$\theta(x) = \sum_{p \leq x} \log p \leq \log x \sum_{p \leq x} 1 = \pi(x) \log x,$$

and applying this and Lemmas 2 and 10, we have

$$\pi(x) \geq \frac{\theta(x)}{\log x} \geq \frac{\psi(x) - x^{1/2} \log^2 x}{\log x} \geq \frac{1}{2}\frac{x}{\log x} - x^{1/2} \log x.$$

Exercise 15 (optional). Verify that $(\log^2 x)/x^{1/2} < .15$ when $x > 400,000$.

From Exercise 15, we have, if $x > 400,000$,

$$\pi(x) \geq \frac{1}{2}\frac{x}{\log x} - .15\frac{x}{\log x} \geq \frac{1}{3}\frac{x}{\log x},$$

and thus the theorem is proved.

We have not been as accurate as we could have been in estimating constants. By refining Tchebyshev's method, it is possible to prove that for x sufficiently large,

$$\frac{.95695x}{\log x} \leq \pi(x) \leq \frac{1.04423x}{\log x}.$$

It is also known that for $x > 400,000$,

$$\frac{.96x}{\log x} \leq \pi(x) \leq \frac{1.12x}{\log x}.$$

But neither of these is as good as the exact statement

$$\lim_{x \to \infty} \frac{\pi(x)}{x/\log x} = 1,$$

which is the Prime Number Theorem.

Problems

1. Let p_n denote the nth prime.
  (a) Observe that $\pi(p_n) = n$.
  (b) Observe that if $n > 400,000$, then $p_n > 400,000$.
  (c) Show that if $n > 400,000$, then

$$\frac{3n}{10} \leq \frac{p_n}{\log p_n} \leq 3n.$$

  (d) Show that if $n > 400,000$, then $\pi(n^2) > n$.
  (e) Let $x_n = p_n/n$. Show that if $n > 400,000$, then $1 < x_n < n$.
  (f) Show that if $n > 400,000$, then

$$\frac{3}{10} \leq \frac{x_n}{\log nx_n} \leq 3.$$

  (g) Show that if $n > 400,000$, then

$$\frac{3}{10}n \log n \leq p_n \leq 6n \log n;$$

thus p_n increases at about the same rate as $n \log n$.

2. Using part (g) of the last problem, show that

$$\sum_{n=1}^{\infty} 1/p_n$$

diverges.

3. Using part (c) of Problem 1, determine the rate of growth of

$$\sum_{n=1}^{N} (\log p_n)/p_n.$$

4. (a) Show that

$$\frac{(x + 1) \log (x + 1)}{x \log x}$$

decreases as x increases.

(b) Hence conclude that for $n > 400{,}000$,

$$(n + 1) \log (n + 1) \le \frac{21}{20} n \log n.$$

(c) From (b) and part (g) of Problem 1, show that

$$p_{n+1} < 21p_n$$

for n sufficiently large. (It is in fact true that $p_{n+1} < 2p_n$ for all n.)

5. (a) Observe that there is a unique integer r_p such that

$$p^{r_p} \le 2n < p^{r_p+1}$$

for any n and each prime p.

(b) Among the integers $1, 2, \ldots, 2n$ what is the highest power of p that divides any one of them?

(c) Show that

$$M(2n) = \prod_{p \le 2n} p^{r_p}.$$

(d) Show that $\psi(2n) = \log M(2n)$.

(e) Show that the highest power of p that divides $\binom{2n}{n}$ is

$$\sum_{m=1}^{r_p} ([2n/p^m] - 2[n/p^m]).$$

(f) From (e), conclude that $\binom{2n}{n} \,\big|\, M(2n)$.

(g) Show that $\binom{2n}{n} \ge 2^n$.

(h) From (f) and (g), conclude that $M(2n) \ge 2^n$.

(i) From (d) and (h), conclude that $\psi(2n) \ge n \log 2$.

(j) From (i), conclude that for $x > 40$, $\psi(x) > x/3$. (Or some similar relation with different constants.)

Miscellaneous Problems

1. Show that $(2^n + (-1)^{n+1})/3$ is an odd integer for $n \geq 1$.

2. A man bought some six-cent stamps, $1/4$ as many five-cent stamps, and some ten-cent stamps for $5.00. How many stamps of each kind did he buy?

3. In 1494, L. Pacioli said that 134217727 is prime. The number is $2^{27} - 1$; show him wrong.

4. If p and q are primes, greater than or equal to 5, show that $24 \mid (p^2 - q^2)$.

5. (a) What can a square be, modulo 9?
   (b) Is 314,159,267,144 a square?

6. Let a' denote the solution of $ax \equiv 1 \pmod{p}$, $a = 1, 2, \ldots, p - 1$.
   (a) Prove that $(ab)' \equiv a'b' \pmod{p}$.
   (b) Disprove that $(a + b)' \equiv a' + b' \pmod{p}$.

7. Use Table A (p. 209) to find the smallest six consecutive odd composite numbers.

8. Show that if
$$a = r^2 - 2rs - s^2,$$
$$b = r^2 + s^2,$$
$$c = r^2 + 2rs - s^2$$

for some integers r, s, then a^2, b^2, c^2 are three squares in arithmetic progression.

9. Pascal once wrote that he had discovered that the difference of the cubes of any two consecutive integers, less one, is six times the sum of all the positive integers less than or equal to the smaller one. Prove him right.

10. If a and b are odd, show that $64 \mid (a^2 - 1)(b^2 - 1)$.

11. Induce a theorem from the following facts:
$$3^2 + 4^2 = 5^2,$$
$$10^2 + 11^2 + 12^2 = 13^2 + 14^2,$$
$$21^2 + 22^2 + 23^2 + 24^2 = 25^2 + 26^2 + 27^2,$$
$$36^2 + 37^2 + 38^2 + 39^2 + 40^2 = 41^2 + 42^2 + 43^2 + 44^2.$$

12. A palindrome is a number that reads the same backwards as forwards, such as 3141413.

   (a) How many two-digit palindromes are there?
   (b) How many three-digit ones?
   (c) How many k-digit ones?

13. Let (a, b, c) denote the greatest common divisor of a, b, and c.

   (a) Show that $(a, b, c) = ((a, b), c)$.
   (b) If $(a, b) = (b, c) = (a, c) = 1$, show that $(a, b, c) = 1$.
   (c) Is the converse of the theorem in (b) true?

14. For which positive integers k does
$$kx \equiv 1 \pmod{k(k + 1)/2}$$
have a solution?

15. If $p \geq 5$ is prime, show that $p^2 + 2$ is composite.

16. The following problem is at least 400 years old: Find the number of men, women, and children in a company of 20 if together they pay $20, each man paying $3, each woman $2, and each child $.50.

17. If a and b are positive integers, let us say that a *divides* b *weakly* (or, that a is a *weak divisor* of b), written $a \int b$, if and only if $p \mid a$ implies $p \mid b$ for primes p.

   (a) Find examples of integers a and b such that $a > b$ and $a \int b$.
   (b) Prove that $a \mid b$ implies $a \int b$.
   (c) Prove that $a \int 1$ implies $a = 1$.
   (d) Prove that $a \int b$ and $b \int c$ implies $a \int c$.
   (e) Prove that $a \int b$ implies $ac \int bc$ for all positive integers c.
   (f) Prove that $a \int b$ and $c \int d$ imply $ac \int bd$.
   (g) Prove that $ab \int c$ implies $a \int c$ and $b \int c$.
   (h) Prove that $ac \int bc$ and $(a, c) \int (b, c)$ imply $a \int b$.
   (i) Prove that $ac \int bc$ and $(a, c) = 1$ imply $a \int b$.
   (j) Prove that $a \int b$ implies $(a, c) \int (b, c)$ for all positive integers c.
   (k) Prove that $a \int b$ and $c \int d$ imply $(a, c) \int (b, d)$.
   (l) Prove that $a \int b$ implies $a^n \int b^m$ for all positive integers m and n.
   (m) Prove that if there are integers m and n such that $a^n \int b^m$, then $a \int b$.
   (n) Prove that $a \int c$ and $b \int c$ imply $ab \int c$.
   (o) Which of (c) to (n) are false for ordinary divisibility of positive integers? Give examples.

18. Construct a formula for f such that $f(n)$ is $1/2$ if n is even and 1 if n is odd.

19. Find the smallest integer n, $n > 0$, such that
$$2 \mid n, 3 \mid n + 1, 5 \mid n + 2, 7 \mid n + 3, 11 \mid n + 4, 13 \mid n + 5.$$
What is n if the condition $17 \mid n + 6$ is added?

20. Show that if $n = a^2 + b^2 = c^2 + d^2$, then

$$n = \frac{((a - c)^2 + (b - d)^2)((a + c)^2 + (b - d)^2)}{4(b - d)^2},$$

and hence that if n can be written as a sum of two squares in two distinct ways, then n is composite.

21. Using the result of Problem 20, factor

(a) $533 = 23^2 + 2^2 = 22^2 + 7^2$,
(b) $1073 = 32^2 + 7^2 = 28^2 + 17^2$.

22. Use Table B and Problem 20 to factor (a) 170,833, (b) 182,410.

23. Show that if $a + b$ is even, then $24 \mid ab(a^2 - b^2)$.

24. If $n = a^2 + b^2 + c^2$, with a, b, c nonnegative, show that

$$(n/3)^{1/2} \leq \max(a, b, c) \leq n^{1/2}.$$

25. Gauss proved that a regular polygon with m sides can be constructed with ruler and compass if $m = 2^a n$, where a is an integer and $n = 1$ or n is a product of distinct primes of the form $2^k + 1$. List all the regular polygons with fewer than 40 sides that can be constructed with ruler and compass.

26. $1^2 + 2^2 = 3^2 - 2^2$. $2^2 + 3^2 = 7^2 - 6^2$. $3^2 + 4^2 = 13^2 - 12^2$. $4^2 + 5^2 = 21^2 - 20^2$. What happens in general?

27. It is known that $8k + 3 = x^2 + y^2 + z^2$ has a solution for any $k \geq 0$.

(a) Show that x, y, and z are odd.
(b) Deduce that k is equal to a sum of three triangular numbers.

28. $(1/3)^2 + (2/3) = (1/3) + (2/3)^2$; is this astonishing?

29. (a) Find a solution of $x^4 + y^4 = z^4$ in real numbers, with $xyz \neq 0$.
(b) Given that $x^4 + y^4 = z^4$ has no solutions in integers, prove that it has no solutions in rational numbers.

30. Verify that $(5 + \frac{5}{24})^{1/2} = 5(\frac{5}{24})^{1/2}$. Are there other numbers like that?

31. Prove

(a) If $a \mid c$, $b \mid c$, and $(a, b) = d$, then $ab \mid cd$.
(b) If $(a, c) = 1$ and $(b, c) = d$, then $(ab, c) = d$.

32. Is $.123456789101112131415\ldots$ rational?

33. (a) Let n be an integer written in the base 12, and let m be its reversal. Show that $\varepsilon \mid (n - m)$.
(b) Generalize to any base b.

34. (a) Suppose that $0 \leq m < 121$. If $210n + m$ is prime, show that m is prime.
(b) Generalize: prove that if $P_k = p_1 p_2 \cdots p_k$ (p_i denotes the ith prime) and $0 \leq m < p_{k+1}^2$, then $P_k n + m$ prime implies m prime.

35. (a) Find all primes p such that $3p + 1$ is a square.
(b) Find all primes p such that $3p + 2$ is a square.

36. Fermat was sometimes as blind as the rest of us. He wrote to Roberval, "Permit me to ask you for the demonstration of this proposition which I frankly confess I have not yet been able to find, although I am assured that it is true. If a, b are integers, and if

(1) $$a^2 + b^2 = 2(a + b)x + x^2,$$

both x and x^2 are irrational." Help Fermat out: show that the equation is not satisfied for any integer x, unless $a = b = 0$.

37. Apply the rational root theorem to complete the demonstration in Problem 36. That is, show that if (1) has no integer roots, then it has no rational roots. What conditions on a and b guarantee that x^2 is irrational?

38. If $n = (6m + 1)(12m + 1)(18m + 1)$, show that $n - 1$ is divisible by $36m$.

39. Several years ago today, a man borrowed an integer number of dollars at a normal rate of simple interest. Today he repaid the loan in full with \$204.13. How much did he borrow, how long ago, and at what rate of interest?

40. (a) Show that $111 \ldots 11$ (n digits, all ones) is composite if n is composite.
(b) Is the converse true?

41. Prove that $a \mid c$, $c \mid b$, $(a, b) = 1$ implies $a = \pm 1$.

42. Which of $2 + 1$, $2 \cdot 3 + 1$, $2 \cdot 3 \cdot 5 + 1$, $2 \cdot 3 \cdot 5 \cdot 7 + 1$, $\ldots$ can be written as a sum of two squares?

43. If n is an even perfect number, $n \neq 6$, show that the last digit in its duodecimal representation is four.

44. Let $f(x)$ be nonnegative and nondecreasing for $x \geq 0$. Let us say that f is *sort of multiplicative* (*smult* for short) if and only if $f(nm) \geq f(n)f(m)$ for all positive integers m and n.

(a) Show that $f(n) = n^k$, k a positive integer, is smult.
(b) Prove that the product of two smult functions is smult.
(c) Show that if $g(x)$ is nonnegative for $x \geq 0$, then $n^{g(n)}$ is smult.

45. Let $f(n)$ denote the number of positive odd divisors of n.

(a) Make a table of f for $n = 2, 3, 4, \ldots, 15$.
(b) Show that $f(2^n p^m) = m + 1$ (p an odd prime).
(c) Guess a formula for $f(2^n p_1^{e_1} p_2^{e_2} \cdots p_k^{e_k})$ (p_i an odd prime).
(d) Prove it by induction on k.

46. Show that
$$2! \, 4! \cdots (2n)! \geq ((n + 1)!)^n$$
for $n = 1, 2, \ldots$.

47. (a) Prove that $(a, b) = (a, c) = 1$ implies $(a, bc) = 1$.
(b) Prove that $(a, b) = 1$ implies $(a^n, b^m) = 1$ for any positive integers m and n.

48. Prove that the sum of twin primes (that is, $2p + 2$, where p and $p + 2$ are both primes) is divisible by 12 if $p > 3$.

49. If $p \mid (ra - b)$ and $p \mid (rc - d)$, prove that $p \mid (ad - bc)$.

50. If $d > 0$, $d \mid n$ and $(d, n/d) = 1$, then d is called a *unitary divisor* of n.

(a) What are the unitary divisors of 120? Of 360?
(b) Which integers are such that their only divisors are unitary divisors?

(c) If $n = p_1^{e_1} p_2^{e_2} \cdots p_k^{e_k}$, how many unitary divisors has n?

(d) If
$$\sum d = 2n,$$

where the sum is taken over the unitary divisors, d, of n, then n is called a *unitary perfect* number. Find two such numbers.

51. Show that every $n > 0$ satisfies at least one of

$$n \equiv 0 \ (\text{mod } 2) \qquad n \equiv 0 \ (\text{mod } 3) \qquad n \equiv 1 \ (\text{mod } 4)$$
$$n \equiv 3 \ (\text{mod } 8) \qquad n \equiv 7 \ (\text{mod } 12) \qquad n \equiv 23 \ (\text{mod } 24).$$

52. Show that 99^9 ends in 89.

53. If p is a prime and $ap + b = c^2$, then show that all values of k making $kp + b$ a square are given by

$$k = pn^2 \pm 2cn + a,$$

where n is any integer.

54. If $m > 1$ is odd, show that $2^m + 1$ is composite.

55. Prove by induction that $3^{n+2} \mid 10^{3^n} - 1, \ n = 0, 1, 2, \ldots$.

56. Let us call n a *practical number* if every positive integer less than or equal to n is a sum of distinct divisors of n.

(a) Show that 12 is practical.

(b) Show that 10 is not.

(c) Discover a practical number greater than 12.

(d) Show that every power of two is practical.

(e) Show that every even perfect number is practical.

57. Show that $x^6 + 2^6 = z^6$ has no solutions with $(x, z) = 1$.

58. (a) Show that if $p \mid (n^2 + 2an + b)$ for some n, then $((a^2 - b)/p) = 1$.

(b) Which primes can divide $n^2 + 2n + 2$?

59. Suppose that we have a solution of

$$ab(a + b)(a - b) = c^2$$

where a, b, and c have no common factors.

(a) Show that a and b are both odd.

(b) Show that any two of a, b, $(a + b)/2$, and $(a - b)/2$ are relatively prime.

(c) Conclude that each of a, b, $(a + b)/2$, and $(a - b)/2$ is a square.

(d) Put $(a + b)/2 = r^2$ and $(a - b)/2 = s^2$. What are a and b in terms of r and s?

(e) Conclude that if there is a solution of $ab(a^2 - b^2) = c^2$ where a, b, c have no common factor, then there is a solution of

$$r^2 + s^2 = t^2,$$
$$r^2 - s^2 = u^2.$$

60. Show that there are infinitely many square triangular numbers, and find four such squares.

61. Show that every positive integer n can be written

$$n = x^2 + y^2 - z^2$$

for some integers x, y, z.

62. Let $\langle x \rangle$ denote the fractional part of x. That is, $\langle x \rangle = x - [x]$.

(a) Show that if $(a, n) = 1$, then the set of numbers

$$\langle a/n \rangle, \langle 2a/n \rangle, \ldots, \langle (n-1)a/n \rangle$$

is a permutation of the set

$$1/n, 2/n, \ldots, (n-1)/n.$$

(b) Show that if $(a, n) = 1$, then

$$\sum_{k=0}^{n-1} \left[\frac{ak}{n} \right] = \frac{(a-1)(n-1)}{2}.$$

63. Show that not every positive integer n can be written $n = x^2 - y^2$ for some integers x, y.

64. Show that

$$x^3 + y^3 + z^3 + x^2y + y^2z + z^2x + xyz = 0$$

has no nontrivial solutions.

65. Find m and n such that

$$(m, n), \qquad (m+1, n), \qquad (m, n+1), \qquad (m+1, n+1)$$

are all greater than 1.

66. (a) Show that $n^2 + (n+1)^2 + (n+2)^2 = m^2$ is impossible.

(b) Show that $n^2 + (n+1)^2 + \cdots + (n+k)^2 = m^2$ is impossible whenever $1^2 + 2^2 + \cdots + k^2$ is a quadratic nonresidue (mod $k+1$).

(c) What are the first three such values of k?

67. Show that the last nonzero digit of $n!$ is even when $n > 1$.

68. Show that no power of 2 is a sum of two or more consecutive positive integers.

69. Let $f(n)$ denote the smallest positive integer m such that $m! \equiv 0 \pmod{n}$.

(a) Make a table of f for $n = 2, 3, \ldots, 20$.

(b) Show that $f(p) = p$.

(c) Show that if p and q are distinct primes, then $f(pq) = \max(p, q)$.

(d) Show that if $p > k$, then $f(p^k) = kp$.

70. Show that

$$\sum_{k=1}^{n} k!$$

is never a square when $n > 3$.

71. Find nine integers in arithmetic progression whose sum of squares is a square.

72. (a) What is the largest integer with ten distinct digits that is divisible by 9?

(b) What is the largest integer with eight distinct digits that is divisible by 9?

(c) What is the largest integer divisible by 11 whose four distinct digits are 1, 2, 3, 4?

(d) What is the largest integer divisible by 11 whose nine distinct digits are 1, 2, . . . , 9?

73. If n is composite, show that $\phi(n) \leq n - n^{1/2}$.

74. How many solutions has $n = xy$ with $(x, y) = 1$?

75. Write the positive integers in a spiral-like array as shown:

$$
\begin{array}{ccccc}
17 & 16 & 15 & 14 & 13 \\
18 & 5 & 4 & 3 & 12 \\
19 & 6 & 1 & 2 & 11 \\
20 & 7 & 8 & 9 & 10 \\
21 & 22 & 23 & \cdots &
\end{array}
$$

If 1 is at the origin of a rectangular coordinate system and $n \geq 0$,

(a) What integer is at $(n, 0)$?
(b) What integer is at (n, n)?
(c) What integer is at $(-n, 0)$?
(d) Where is $(2n + 1)^2$?
(e) Where is $(2n)^2$?
(f) Where is 1000?

76. If p is a prime, show that $p + 2$ is prime if and only if

$$4((p - 1)! + 1) + p \equiv 0 \ (\text{mod } p + 2).$$

77. (a) Given $n > 0$, show that there is an integer m such that

$$((n + 1)^{1/2} + n^{1/2})^2 = (m + 1)^{1/2} + m^{1/2}.$$

(b) Can you always find an m such that

$$((n + 1)^{1/2} + n^{1/2})^3 = (m + 1)^{1/2} + m^{1/2}?$$

78. True or false? If $(m, n) = 1$, then the product of the $\phi(n)$ numbers less than or equal to n and prime to n with the $\phi(m)$ numbers less than or equal to m and prime to m give the $\phi(mn)$ numbers less than or equal to mn and prime to mn.

79. If $(a, m) = 1$, then the least residues (mod m) of $a, 2a, \ldots, ma$ are a permutation of $1, 2, \ldots, m$. What happens if $(a, m) \neq 1$?

80. Let p_i denote the ith prime. Show that $P_n = p_1 p_2 \cdots p_n + 1$ is never a square.

81. Which positive integers are neither composite nor the sum of two positive composite integers?

82. If p and q are primes and $p > q$, show that $\phi(p^a) = \phi(q^b)$ implies $a = 1$. ($a > 0$ and $b > 0$.)

83.
$$
\begin{aligned}
2^2(2^3 - 1) &= 1^3 + 3^3; \\
2^4(2^5 - 1) &= 1^3 + 3^3 + 5^3 + 7^3; \\
2^6(2^7 - 1) &= 1^3 + 3^3 + \cdots + 15^3.
\end{aligned}
$$

We might induce that every even perfect number, except 6, is a sum of consecutive odd cubes, starting with 1^3. Is this so?

84. $1000 \equiv 1 \pmod{37}$. From this, develop a test for the divisibility of an integer by 37.

85. (a) Find all x, y, z in arithmetic progression such that
$$x^2 + xy + y^2 = z^2$$
and $xy \neq 0$.

(b) Find all x, y, z in arithmetic progression such that
$$x^2 + kxy + y^2 = z^2$$
and $xy \neq 0$.

86. If a and b are odd, show that there is a solution of
$$x^2 + y^2 = (a^2 + b^2)/2.$$

87. If n is an even perfect number, show that the harmonic mean of the divisors of n is an integer.

88. Suppose that $n^2 + n + 1 = p^r$, where $n \geq 1$, p is prime, and $r \geq 1$.

(a) Show that p is odd.
(b) Show that if $n \equiv 1 \pmod 3$, then the only solution is $n = 1, p = 3, r = 1$.
(c) Show that r is odd.
(d) Show that if $p \neq 3$, then $p \equiv 1 \pmod 3$.

89. Let $f(x) = a_n x^n + a_{n-1} x^{n-1} + \cdots + a_0$. Suppose that a_n, a_0, and an odd number of the remaining coefficients are odd. Show that $f(x) = 0$ has no rational roots.

90. If $n < p$ and $(n, p - 1) = 1$, show that $x^n \equiv n \pmod p$ has a solution.

91. (a) Show that if $(a, p) = 1$, $n \mid (p - 1)$, and $a^{(p-1)/n} \not\equiv 1 \pmod p$, then a is not an nth power residue $\pmod p$.

(b) Is 2 a fifth power residue $\pmod{31}$?

92. Show that $n \mid (2^n + 1)$ if n is a power of 3.

93. (a) Show that $n^2 + (n + 1)^2 = 3m^2$ is impossible.

(b) What is a necessary condition for
$$n^2 + (n + 1)^2 = km^2$$
to have a solution for given k, $k > 0$?

94. Let m be square-free. (That is, $m = p_1 p_2 \cdots p_k$, a product of *distinct* primes.) Suppose that m has the property that $p \mid m$ implies $(p - 1) \mid m$.

(a) Show that $m = 2$, 6, and 42 have this property.
(b) Does any other m?
(c) How many others?

95. A man sold n cows for $\$n$ per cow. With the proceeds, he bought an odd number of sheep for $10 each and a pig for less than $10. How much did the pig cost?

96. Show that
$$\prod_{n < p \leq 2n} p \;\Big|\; \frac{(2n)!}{n!n!}.$$

97. Prove that if p and $q = 6p + 1$ are odd primes, then 3 is a primitive root of q.

98. If $p = 2^m + 1$ is a prime, show that every quadratic nonresidue of p is a primitive root of p.

99. Solve $x(x - 31) = y(y - 41)$ in positive integers.

100. Let c_0, c_1, c_2, c_3, c_4 be any integers. Show that

$$f(x) = c_0 + c_1 \binom{x}{1} + 2c_2 \binom{x}{2} + 6c_3 \binom{x}{3} + 12c_4 \binom{x}{4}$$

has the property that, given m,

$$m \mid f(n + m) - f(n)$$

for all integers n. $\left(\text{For the notation } \binom{x}{r} \text{ see Appendix B, p. 196.}\right)$

101. Here is a proof that the sequence $\{2kn + 1\}$, $k = 1, 2, \ldots$ contains infinitely many primes. Fill in any missing details. If $p_1, p_2, \ldots, p_n$ are all of the primes of the form $2kn + 1$ and $N = p_1 p_2 \cdots p_n$, then $N^n + 1$ is divisible by none of them. Since $(N^n + 1)/(N + 1)$ has no prime divisors not of the form $2kn + 1$, there exists at least one greater than p_n.

102. Show that

$$\sum_{n=1}^{N} \sum_{d \mid n} f(d, n) = \sum_{m=1}^{N} \sum_{r=1}^{[N/m]} f(m, rm).$$

103. (a) Show that if r and s satisfy $5^n s - 2^n r = 1$ and $x = 5^n s$, then the last n digits of x^2 are the same as the last n digits of x.
(b) Find such a number for $n = 3$.

104. Establish the following test for primes: if α is a primitive nth root of 1 (that is, $\alpha^n = 1$ and $\alpha^m \neq 1$ for $0 < m < n$), then

$$\frac{1}{n} \sum_{j=0}^{n-1} \alpha^{j(1+(n-1)!)} = \begin{cases} 1 & \text{if } n \text{ is prime} \\ 0 & \text{if } n \text{ is composite.} \end{cases}$$

105. Show that the sequence $\{2 + np\}$, p an odd prime, $n = 1, 2, \ldots$, contains an infinite geometric progression for any p.

Bibliography

Number theory has a huge literature. Books on the subject have been printed for almost 400 years, and the last word has yet to be said. The following list of titles is certainly not exhaustive; it contains purely personal suggestions for further reading. Books marked with an asterisk (*) are available in paperback.

The following seven textbooks are arranged, roughly, with the easiest listed first and the more difficult last. They are all superior books; those by Ore and Niven and Zukerman are especially good.

Oystein Ore, *Number Theory and Its History*, New York: McGraw-Hill, 1948.

B. W. Jones, *The Theory of Numbers*, New York: Holt, 1955.

J. V. Uspensky and M. A. Heaslet, *Elementary Number Theory*, New York: McGraw-Hill, 1939.

*Harriet Griffin, *Elementary Theory of Numbers*, New York: McGraw-Hill, 1954.

W. J. LeVeque, *Topics in Number Theory* (vol. 1), Reading, Mass.: Addison-Wesley, 1956.

Ivan Niven and H. S. Zukerman, *An Introduction to the Theory of Numbers*, New York: Wiley, second edition, 1966.

Trygve Nagell, *Introduction to Number Theory*, Bronx, N. Y.: Chelsea, second edition, 1964.

The best book for further browsing in number theory is, I think,

G. H. Hardy and E. M. Wright, *The Theory of Numbers*, New York: Oxford Univ. Press, fourth edition, 1960.

There is no better place to get the flavor of number theory than in this elegantly written book. Though not a text, it does include introductions to many branches of number theory. The subjects it contains range widely in level of difficulty—from how to win at the game of Nim to a proof of the Prime Number Theorem.

Hardy was a great number theorist; his views on mathematics and its relation to life are given in a short nontechnical book that is well worth reading:

> G. H. Hardy, *A Mathematician's Apology*, New York: Cambridge Univ. Press, revised edition, 1967.

Two unusual texts that contain interesting material are

> Waclaw Sierpinski, *Elementary Theory of Numbers*, New York: Hafner, 1964.
>
> Daniel Shanks, *Solved and Unsolved Problems in Number Theory*, New York: Spartan, 1962.

Noteworthy among popular works are

> *P. J. Davis, *The Lore of Large Numbers*, New York: Random House, 1961.
>
> *Ivan Niven, *Numbers*, New York: Random House, 1961.
>
> *A. H. Beiler, *Recreations in the Theory of Numbers*, New York: Dover, 1964.
>
> *Tobias Dantzig, *Number*, New York: Macmillan, fourth edition, 1967.

The first two mix exposition with information in a pleasant and elementary manner. The author of the third is crazy about numbers; he reports on various results and calculations with great enthusiasm. The fourth is mostly historical and philosophical.

Six specialized books conclude this bibliography. The first two are reference works, and the other four are well-written introductions to some topics of number theory.

> L. E. Dickson, *History of the Theory of Numbers* (3 vols.), Bronx, N. Y.: Chelsea (reprint of the 1919 edition).

Dickson's work attempts to give a short description of *everything* published in number theory from the beginning of time until 1918. It is thus invaluable for reference, and if ever you make what you think is a new discovery in number theory, look in Dickson: it may be there, discovered by someone else.

> D. N. Lehmer, *Factor Table for the First Ten Millions*, New York: Hafner (reprint of the 1909 edition).

There is no danger that Lehmer's work will become obsolete.

Waclaw Sierpinski, *Pythagorean Triangles*, New York: Academic Press, 1962.

*A. O. Gelfond, *The Solution of Equations in Integers*, San Francisco: W. H. Freeman and Company, 1961.

Ivan Niven, *Irrational Numbers*, New York: Wiley, 1956.

Harry Pollard, *The Theory of Algebraic Numbers*, New York: Wiley, 1950.

Proof by Induction

In the text, the method of proof by mathematical induction is used several times. The purpose of this section is to recall what the method is, show some examples of how it operates, and give some problems for practice.

It is notorious that mathematics is a deductive art: starting with a collection of postulates, theorems are deduced by following the laws of logic. That is the way it is presented in print, but it is not the way that most new mathematics is discovered. It is difficult to sit down and think, "I will now deduce," and prove anything worthwhile. The goal must be in sight; you must suspect that a theorem is true, and then deduce it from what you know. The theorem you suspect is true must come from somewhere: many theorems are the result of correct guesses.

Exercise 1. Guess what $f(n)$ is from the following data:

n	0	1	2	3	4	5
$f(n)$	1	0	1	4	9	16.

Exercise 2. Guess what $f(n)$ is from

n	0	1	2	3	4	5	6
$f(n)$	1	2	5	10	17	26	37.

Exercise 3 (optional). Guess a theorem about $f(n)$:

n	1	2	3	4	5	6	7
$f(n)$	2	-1	-2	-1	2	7	14.

Since number theory is largely concerned with the positive integers, some of its theorems are of the form, "Such-and-such is true for all positive integers *n.*" Propositions like this can often be proved by mathematical induction (or induction for short—we will not be concerned with any other kind). This method of proof is based on the following property of positive integers:

(1)
> If a set of integers contains 1, and
> if it contains $r + 1$ whenever it contains r,
> then the set contains all the positive integers.

(This property is so fundamental that it is usually taken as a nonprovable *postulate* about the positive integers.) It is applied when we want to show that a proposition $P(n)$ about the positive integer n is true for all $n, n = 1, 2, \ldots$. Examples of such propositions are

$P_1(n)$: "$n^2 + 3n + 2 > (n + 1)^2 - 5$."

$P_2(n)$: "$n(n + 1)(n + 2)$ is divisible by 6."

$P_3(n)$: "$f(x_1 + x_2 + \cdots + x_n) \geq f(x_1) + f(x_2) + \cdots + f(x_n)$."

Let S denote the set of positive integers for which $P(n)$ is true. If we can show that 1 is in S and that if r is in S, then $r + 1$ is in S, then (1) says that all positive integers are in S. Rephrasing this, we get the induction principle:

> If $P(1)$ is true, and
> if the truth of $P(r)$ implies the truth of $P(r + 1)$,
> then $P(n)$ is true for all $n, n = 1, 2, \ldots$.

Exercise 4. Fill in the blank: if $P(17)$ is true, and if the truth of $P(r)$ implies the truth of $P(r + 1)$, then $P(n)$ is true for _____.

To illustrate a proof by induction, we will take a well-known example. Let $P(n)$ be the statement

$$\text{"}1 + 2 + \cdots + n = n(n + 1)/2.\text{"}$$

We will prove by induction that $P(n)$ is true for all positive integers n.

Exercise 5. What is $P(1)$? Is it true?

Suppose that $P(r)$ is true; that is, suppose that

(2) $1 + 2 + \cdots + r = r(r + 1)/2.$

We wish to deduce that $P(r + 1)$ is true. That is, we want to show that

(3) $1 + 2 + \cdots + (r + 1) = (r + 1)(r + 2)/2$

follows from (2). If we add $r + 1$ to both sides of (2), we get

$$1 + 2 + \cdots + r + (r + 1) = r(r + 1)/2 + (r + 1)$$

$$= (r + 1)\left(\frac{r}{2} + 1\right) = (r + 1)(r + 2)/2,$$

which is (3). Hence both parts of the induction principle have been verified, and it follows that $P(n)$ is true for all positive integers n.

In any proof by induction, we must not forget to show that $P(1)$ is true. Even if we show that the truth of $P(r)$ implies the truth of $P(r + 1)$, if $P(1)$ is not true, then we cannot conclude that $P(n)$ is true for any n. For example, let $P(n)$ be

$$n + (n + 1) = 2n.$$

Suppose that $P(r)$ is true. That is, we assume that

(4) $$r + (r + 1) = 2r.$$

Using this, we have

$$(r + 1) + (r + 2) = r + (r + 1) + 2 = 2r + 2 = 2(r + 1),$$

so $P(r + 1)$ is true. So, *if* $P(1)$ were true, it would follow that $P(n)$ is true for all positive integers n. Since $P(1)$ is not true, we cannot so conclude. In fact, $P(n)$ is false for all n.

It should go without saying that in any proof by induction, we must verify that the truth of $P(r)$ implies the truth of $P(r + 1)$. For example, from the table

n	1	2	3	4	5	6
$f(n)$	2	4	6	8	10	12

we cannot conclude that $f(n) = 2n$ for all n. In fact, $f(7) = \pi$, because the function that I had in mind when constructing the table was

$$f(n) = 2n + \frac{(n - 1)(n - 2)(n - 3)(n - 4)(n - 5)(n - 6)(\pi - 14)}{6 \cdot 5 \cdot 4 \cdot 3 \cdot 2}.$$

Another form of the induction principle is sometimes used:

If $P(1)$ is true, and
if the truth of $P(k)$ for $1 \le k \le r$ implies the truth of $P(r + 1)$,
then $P(n)$ is true for all n, $n = 1, 2, \ldots$.

This is valid because of the corresponding property of integers: if a set of integers contains 1, and contains $r + 1$ whenever it contains $1, 2, \ldots, r$, then it contains all positive integers.

Problems

1. Prove that
$$1/1\cdot2 + 1/2\cdot3 + \cdots + 1/(n-1)n = 1 - (1/n)$$
for $n = 2, 3, \ldots$.

2. 1, 3, 6, and 10 are called *triangular numbers:*

Let t_n denote the nth triangular number. Find a formula for t_n.

3. Prove that
$$1^2 + 2^2 + \cdots + n^2 = n(2n+1)(n+1)/6$$
for $n = 1, 2, \ldots$.

4.
$$1^3 = 1^2,$$
$$1^3 + 2^3 = 3^2,$$
$$1^3 + 2^3 + 3^3 = 6^2,$$
$$1^3 + 2^3 + 3^3 + 4^3 = 10^2.$$

Guess a theorem and prove it.

5. From Problem 4, or by guessing and induction, derive a formula for
$$1^3 + 3^3 + 5^3 + \cdots + (2k-1)^3,$$
$k = 1, 2, \ldots$.

6. Suppose that $a_1 = 1$ and $a_{n+1} = 2a_n + 1$, $n = 1, 2, \ldots$. Prove by induction that $a_n = 2^n - 1$.

7. Suppose that $a_0 = a_1 = 1$ and $a_{n+1} = a_n + 2a_{n-1}$, $n = 1, 2, \ldots$. Prove by induction that
$$a_n = \frac{2^{n+1} + (-1)^n}{3}.$$

8.
$$1\cdot2\cdot3\cdot4 = 5^2 - 1,$$
$$2\cdot3\cdot4\cdot5 = 11^2 - 1,$$
$$3\cdot4\cdot5\cdot6 = 19^2 - 1,$$
$$4\cdot5\cdot6\cdot7 = 29^2 - 1.$$

Guess and prove a theorem. (Induction may not be necessary.)

9. Guess and prove a formula for
$$1^2 + 4^2 + 7^2 + \cdots + (3n+1)^2,$$
$n = 0, 1, \ldots$.

10. Prove by induction that $n(n+1)(n+2)$ is divisible by 6 for $n = 1, 2, \ldots$.

11. Construct a formula for a function f such that
$$f(1) = f(2) = f(3) = f(4) = 0, \qquad f(5) = 17.$$

12. Let t_n denote the nth triangular number. Consider the table

n	1	2	3	4	5
t_n	1	3	6	10	15
$8t_n + 1$	9	25	49	81	121.

Are all those squares a coincidence?

13. Prove by induction that $n^5 - n$ is divisible by 5, $n = 1, 2, \ldots$.

14. The *Fibonacci numbers* are defined by
$$f_{n+1} = f_n + f_{n-1}, \qquad f_1 = f_2 = 1.$$

Prove that f_{5n} is divisible by 5, $n = 1, 2, \ldots$.

Summation and Other Notations

The summation sign $\sum$ is very helpful indeed. As it is usually used, the variable of summation assumes all the integer values in the indicated range:

$$\sum_{i=3}^{9} i^2 = 9 + 16 + 25 + 36 + 49 + 64 + 81,$$

$$\sum_{j=2}^{k} n/j = \frac{n}{2} + \frac{n}{3} + \cdots + \frac{n}{k},$$

$$\sum_{k=1}^{n} kt_k = t_1 + 2t_2 + 3t_3 + \cdots + nt_n,$$

$$\sum_{r=1}^{5} 1 = 1 + 1 + 1 + 1 + 1.$$

Exercise 1. Write out

(a) $\displaystyle\sum_{r=1}^{6} f(r)g(7 - r),$ (b) $\displaystyle\sum_{k=1}^{3} \varphi(k)/k,$ (c) $\displaystyle\sum_{k=-1}^{1} \sum_{j=2}^{4} f(j)g(k).$

Exercise 2. Write in summation notation

(a) $1 + 2 + 3 + \cdots + k,$ (b) $\frac{1}{3} + \frac{2}{5} + \frac{3}{7} + \cdots + \frac{17}{35},$
(c) $2 + \frac{3}{4} + \frac{4}{9} + \cdots + \frac{17}{256}.$

In general, $\sum_I f(a)$ tells us to add up the values of $f(a)$, following the instructions I, which determine a set of values for a. For example,

$$\sum_{d|n} f(d)$$

denotes a sum extended over the positive divisors of n:

$$\sum_{d|9} d^2 = 1^2 + 3^2 + 9^2,$$

$$\sum_{d|12} \sigma(d) = \sigma(1) + \sigma(2) + \sigma(3) + \sigma(4) + \sigma(6) + \sigma(12).$$

Exercise 3. Evaluate

(a) $\displaystyle\sum_{d|6} 1/d,$ (b) $\displaystyle\sum_{d|28} 1/3,$ (c) $\displaystyle\sum_{d|15} 15/d.$

Another kind of sum often seen is

$$\sum_{p \le x} f(p),$$

a sum extended over the primes no greater than x. For example,

$$\sum_{p \le 10} f(p) = f(2) + f(3) + f(5) + f(7).$$

Thus, the important function $\pi(x)$, the number of primes less than or equal to x, can be written

$$\pi(x) = \sum_{p \le x} 1.$$

In fact, we can put any sort of condition we please under or around a summation sign. For example,

$$\sum_{\substack{(n,10)=1 \\ n \le 10}} n = 1 + 3 + 7 + 9,$$

$$\sum_{\substack{2 \le m \le 6 \\ 2 \le n \le 6 \\ (m,n)=1}} \frac{m}{n} = \frac{2}{3} + \frac{2}{5} + \frac{3}{2} + \frac{3}{4} + \frac{3}{5} + \frac{4}{3} + \frac{4}{5} + \frac{5}{2} + \frac{5}{3} + \frac{5}{4} + \frac{5}{6} + \frac{6}{5},$$

$$\sum_{\substack{m+n=6 \\ m,n \ge 1}} f(m, n) = f(1, 5) + f(2, 4) + f(3, 3) + f(4, 2) + f(5, 1).$$

Exercise 4. Calculate

(a) $\displaystyle\sum_{p \le 15} p,$ (b) $\displaystyle\sum_{\substack{5|n \\ 0 < n < 49}} \frac{n}{5},$ (c) $\displaystyle\sum_{0 < n^{1/2} \le 99} 1.$

Just as $\sum$ stands for sum, $\prod$ stands for product, and whatever can be done with one notation can be done with the other. For example,

$$\prod_{k=1}^{n} k = n!, \qquad \prod_{p \leq 7} \left(1 - \frac{1}{p}\right) = \frac{1}{2} \cdot \frac{2}{3} \cdot \frac{4}{5} \cdot \frac{6}{7},$$

$$10 \prod_{p \mid 10} \left(1 - \frac{1}{p}\right) = 10 \cdot \frac{1}{2} \cdot \frac{4}{5}, \qquad \prod_{k=1}^{4} \frac{k+2}{k} = \frac{3}{1} \cdot \frac{4}{2} \cdot \frac{5}{3} \cdot \frac{6}{4}.$$

Exercise 5. Evaluate

(a) $\displaystyle\prod_{k=1}^{17} 1,$ (b) $\displaystyle\prod_{k=1}^{17} 2,$ (c) $\displaystyle\prod_{p \leq 5} \frac{p}{p+2}.$

Another notation with which it is good to be familiar is the greatest-integer notation: $[x]$ denotes the greatest integer not greater than x. (*Note:* In this section, x and y will not necessarily denote integers—they may assume any real value. Other lower-case italic letters will still be reserved for integers.) Rephrased, the definition says that $[x]$ is the unique integer satisfying

$$x - 1 < [x] \leq x.$$

Rephrased again, to find $[x]$, locate x on the real axis and go to the left until you come to an integer—that is $[x]$. Rephrased yet again, $[x]$ is the unique integer satisfying

$$[x] \leq x < [x] + 1.$$

For example,

$$[2] = 2, \qquad [5/2] = 2, \qquad [\pi] = 3, \qquad [-1/2] = -1, \qquad [-\pi] = -4.$$

The notation $[x]$ can be read in various ways: "integral part of x," which is common and good, and "square brackets x," which is also common, but not as good.

Exercise 6. Prove that $[x + 1] = [x] + 1$. Can you generalize this?

Exercise 7. Show that none of the following is true for all x and y by constructing counterexamples:

$$[x + y] = [x] + [y], \qquad [x/y] = [x]/[y], \qquad [xy] = [x][y].$$

As an example of the use of the greatest-integer notation, we prove

THEOREM 1. *The number of positive multiples of b less than or equal to a is $[a/b]$.*

PROOF. Let all the positive multiples of b less than or equal to a be

$$b, 2b, 3b, \ldots, kb.$$

Since they are all smaller than or equal to a, we have $kb \le a$ or $k \le a/b$. On the other hand, the next multiple of b after kb is larger than a:

$$(k + 1)b > a \quad \text{or} \quad k > \frac{a}{b} - 1.$$

Thus k, the number of positive multiples of b less than or equal to a, is an integer satisfying

$$\frac{a}{b} - 1 < k \le \frac{a}{b}.$$

This is an interval of length one, so it contains only one integer: it is $[a/b]$. Thus $k = [a/b]$ and the theorem is proved.

Exercise 8. How many integers between 1 and 1977 are divisible by 11?

Another notation connected with $[x]$ is the fractional-part notation: the *fractional part* of x, written $\langle x \rangle$, is defined by

$$\langle x \rangle = x - [x].$$

You can see immediately that $0 \le \langle x \rangle < 1$; other properties are left to the reader to discover.

We will mention two other notations that are used in the text and may be unfamiliar. The notation $n!$ (read "n factorial") is defined for positive integers and denotes the product of the positive integers from 1 to n. That is,

$$n! = n(n - 1)(n - 2) \cdots 3 \cdot 2 \cdot 1 \quad \text{for} \quad n = 1, 2, \ldots .$$

For example, $1! = 1$, $4! = 24$, and $6! = 720$. In addition, $0!$ is defined to be 1.

Connected with the factorial notation are the *binomial coefficients* (also called *combinatorial symbols*) $\binom{n}{m}$. For $n \ge m \ge 0$, these are defined by

$$\binom{n}{m} = \frac{n!}{m!(n - m)!} = \frac{n(n - 1) \cdots (n - m + 1)}{m(m - 1) \cdots 1}.$$

For example,

$$\binom{5}{3} = \frac{5!}{3!2!} = \frac{5 \cdot 4}{2 \cdot 1} = 10,$$

$$\binom{10}{4} = \frac{10!}{4!6!} = \frac{10 \cdot 9 \cdot 8 \cdot 7}{4 \cdot 3 \cdot 2 \cdot 1} = 210.$$

Binomial coefficients have many useful properties. For example, if p is a prime, then $p \mid \binom{p}{n}$ for $n = 2, 3, \ldots, p - 1$. Their major occurence is in the *binomial formula*: for any real numbers x and y (integers or not) and any positive integer n,

$$(x + y)^n = \binom{n}{0} x^n + \binom{n}{1} x^{n-1}y + \binom{n}{2} x^{n-2}y^2$$

$$+ \cdots + \binom{n}{n-1} xy^{n-1} + \binom{n}{n} y^n$$

$$= \sum_{k=0}^{n} \binom{n}{k} x^{n-k}y^k.$$

Problems

1. Calculate

(a) $\displaystyle\sum_{k=1}^{6} (k^2 + k),$ (b) $\displaystyle\sum_{k=1}^{\infty} 1/2^k,$ (c) $\displaystyle\sum_{d \mid 28} 1/d,$

(d) $\displaystyle\sum_{\substack{k=1 \\ (k,3)=1}}^{9} k,$ (e) $\displaystyle\sum_{0 \leq k \leq 15} [k^{1/2}],$ (f) $\displaystyle\sum_{p \leq 50} p.$

2. Write in summation notation

(a) $1 + 9 + 25 + 49 + \cdots + 169,$

(b) $\dfrac{1}{n} + \dfrac{1}{n+1} + \dfrac{1}{n+2} + \cdots + \dfrac{1}{2n},$

(c) $a_0 + a_1 x + a_2 x^2 + \cdots + a_n x^n,$

(d) $u_r v_{n-r} + u_{r-1} v_{n-r+1} + \cdots + u_0 v_n.$

3. Prove or disprove

(a) $\displaystyle\sum_{k=1}^{n} (a_k + b_k) = \sum_{k=1}^{n} a_k + \sum_{k=1}^{n} b_k,$

(b) $\displaystyle\sum_{k=1}^{n} ca_k = c \sum_{k=1}^{n} a_k,$

(c) $\displaystyle\sum_{k=1}^{n} a_k b_k = \sum_{k=1}^{n} a_k \sum_{k=1}^{n} b_k.$

4. Evaluate

(a) $(3^{1/2})^{[2^{1/2}]},$ (b) $[3^{1/2}]^{(2^{1/2})},$

(c) $\displaystyle\prod_{k=1}^{6} [6/k]!,$ (d) $\displaystyle\sum_{r=1}^{3} \sum_{k=1}^{r} k^r,$

(e) $\displaystyle\sum_{k=1}^{3} \sum_{r=1}^{k} k^r.$

5. Evaluate (a) $8!$,   (b) $\binom{8}{4}$,   (c) $\binom{12}{5}$.

6. Prove that

$$\binom{n}{m} = \binom{n}{n-m}$$

for integers n and m with $n \geq m \geq 0$.

7. Use the binomial formula to expand

(a) $(x + 2)^6$,   (b) $(1 - a)^7$.

8. Use the binomial formula to simplify

(a) $x^3 + 9x^2 + 27x + 27$,

(b) $\binom{n}{0}x^n - \binom{n}{1}x^{n-1} + \binom{n}{2}x^{n-2} - \cdots + (-1)^n\binom{n}{n}$.

9. Calculate $[(3/2)^k]$ for $k = 1, 2, 3,$ and 4.

10. Is it true that

(a) $[-x] = -[x]$ for all real numbers x?

(b) $[n^x] = n^{[x]}$ for all real numbers x and all integers n?

11. Let $\|x\|$ denote the integer nearest to x. Express $\|x\|$ in terms of the greatest-integer notation.

12. What values can $[x] + [-x]$ take on?

13. Given a prime p, evaluate

(a) $\displaystyle\sum_{\substack{k=1\\p\nmid k}}^{p^2} 1$,   (b) $\displaystyle\sum_{\substack{k=1\\p\nmid k}}^{n} 1$.

14. How many multiples of seven are there between 1902 and 2038?

15. Show that

$$\sum_{d|n} f(d) = \sum_{d|n} f(n/d)$$

for all integers n and functions f.

16. Show that it is not true that

$$\sum_{p \leq n} f(p) = \sum_{p \leq n} f(n/p)$$

for all integers n and functions f.

17. Prove that

$$\sum_{i=1}^{N}\sum_{j=1}^{N} f(i,j) = \sum_{j=1}^{N}\sum_{i=1}^{N} f(i,j).$$

18. Prove that

$$\sum_{i=1}^{N}\sum_{j=1}^{i} f(i,j) = \sum_{j=1}^{N}\sum_{i=j}^{N} f(i,j).$$

19. Show that

$$\prod_{k=1}^{n} k^{[n/k]} = \prod_{k=1}^{n} [n/k]!.$$

20. Show that for all $n > 1$,

$$\frac{1}{n} \sum_{d|n} d > \frac{1}{n^2} \sum_{d|n} d^2.$$

21. How many integers are in the interval $x < n \le y$?

22. Show that

$$\sum_{k=0}^{n-1} \langle k/n \rangle = \frac{n-1}{2}.$$

23. Suppose that

$$\sum_{k=1}^{N} 1/n_k = 1$$

and the n_k's are positive odd integers. Show that N is odd.

Quadratic Congruences to Composite Moduli

We have not yet studied the quadratic congruence

$$(1) \qquad\qquad Ax^2 + Bx + C \equiv 0 \,(\mathrm{mod}\; m)$$

when the modulus is composite, though we do know a lot about $x^2 \equiv a$ (mod p). In this section we will see how to solve (1).

We start by considering a special case,

$$(2) \qquad\qquad x^2 \equiv a \,(\mathrm{mod}\; m).$$

(Unlike quadratic congruences (mod p), not all congruences like (1) can be put in this form by completing the square. For example, $3x^2 + x + 1 \equiv 0$ (mod 9) cannot.) Let $p_1^{e_1} p_2^{e_2} \cdots p_k^{e_k}$ be the prime-power decomposition of m. Then every solution of (2) will also satisfy the system

$$(3) \qquad\qquad x^2 \equiv a \,(\mathrm{mod}\; p_k^{e_k}), \qquad k = 1, 2, \ldots, n,$$

and on account of the Chinese Remainder Theorem, the converse is also true. Hence, to solve (2) for any m, it is sufficient to know how to solve

$$(4) \qquad\qquad x^2 \equiv a \,(\mathrm{mod}\; p^e)$$

for all primes p and positive integers e. For example, let us solve $x^2 \equiv 9$ (mod 28); we break it into pieces, solve the pieces, and then put them together again. We first find solutions of

$$(5) \qquad\qquad x^2 \equiv 9 \,(\mathrm{mod}\; 4) \qquad \text{and} \qquad x^2 \equiv 9 \,(\mathrm{mod}\; 7).$$

The first congruence has solutions 1 and 3, and the second has solutions 3 and 4. Thus there are four sets of solutions to (5), namely

(a) $x \equiv 1 \pmod 4$ and $x \equiv 3 \pmod 7$,

(b) $x \equiv 1 \pmod 4$ and $x \equiv 4 \pmod 7$,

(c) $x \equiv 3 \pmod 4$ and $x \equiv 3 \pmod 7$,

(d) $x \equiv 3 \pmod 4$ and $x \equiv 4 \pmod 7$.

Thus there will be four solutions to the original congruence modulo 28. They are (a) 17, (b) 25, (c) 3, (d) 11.

If $m = p_1^{e_1} p_2^{e_2} \cdots p_n^{e_n}$ and $x^2 \equiv a \pmod{p_k^{e_k}}$ had s_k solutions, then there are $s_1 s_2 \cdots s_n$ different sets of solutions to system (3), and hence $x^2 \equiv a \pmod m$ will have that number of solutions. In particular, if any of the congruences in (3) has no solutions, then (2) has no solutions. We will now see how many solutions $x^2 \equiv a \pmod{p^e}$ has, and then how to find them.

LEMMA 1. *If p is an odd prime, $p \nmid a$, and a is a quadratic residue (mod p), then $x^2 \equiv a \pmod{p^e}$ has exactly two solutions for each positive integer e.*

PROOF. We will prove the lemma by induction. It is true for $e = 1$ (see Section 11). Suppose that it is true for $e = k - 1$, $k \geq 2$. First we will show that each solution of $x^2 \equiv a \pmod{p^{k-1}}$ can be used to construct two solutions of $x^2 \equiv a \pmod{p^k}$, and then we will show that these are the only solutions. Let r be a solution of $x^2 \equiv a \pmod{p^{k-1}}$. Then

(6) $$r^2 = a + hp^{k-1}$$

for some integer h. Consider the numbers

$$R_j = r + jp^{k-1}, \quad j = 1, 2, \ldots.$$

We have

$$R_j^2 = r^2 + 2rjp^{k-1} + j^2 p^{2k-2},$$

so

$$R_j^2 = a + hp^{k-1} + 2rjp^{k-1} \equiv a + (h + 2rj)p^{k-1} \pmod{p^k}.$$

If we choose j such that

$$R_j^2 \equiv a \pmod{p^k},$$

we will have a solution of $x^2 \equiv a \pmod{p^k}$, and we can do this if there is an integer j such that

$$h + 2rj \equiv 0 \pmod p.$$

There is, from (6): since $(2r, p) = 1$, the congruence has a unique solution. As an example of this process, starting with $5^2 \equiv 7 \pmod 9$, we will construct a solution of $x^2 \equiv 7 \pmod{27}$. We put

$$R_j = 5 + 9j,$$

so

$$R_j^2 = 25 + 90j + 81j^2 \equiv 25 + 9j \,(\text{mod } 27).$$

We want to pick j so as to make $25 + 9j \equiv 7 \,(\text{mod } 27)$. $j = 1$ does this, and hence 14 is a solution of $x^2 \equiv 7 \,(\text{mod } 27)$.

Exercise 1. Verify that $14^2 \equiv 7 \,(\text{mod } 27)$.

Exercise 2. Find a solution of $x^2 \equiv 7 \,(\text{mod } 81)$.

We have shown that $x^2 \equiv a \,(\text{mod } p^k)$ has a solution; call it s. Then $p^k - s$ is another solution. It remains to show that there are no more. Suppose that t is any solution of $x^2 \equiv a \,(\text{mod } p^k)$. We will now show that $t = s$ or $t = p^k - s$, and this will complete the proof of the lemma. We have $t^2 \equiv s^2 \,(\text{mod } p^k)$, so

$$p^k \mid (t - s)(t + s).$$

There are three cases: since $k \geq 2$, either

$$(i) \; p^k \mid (t - s),$$
$$(ii) \; p^k \mid (t + s),$$
$$(iii) \; p \mid (t - s) \quad \text{and} \quad p \mid (t + s).$$

In the first case $t \equiv s \,(\text{mod } p^k)$, and since both s and t are least residues $(\text{mod } p^k)$, we have $t = s$. In the second case, we have $t = p^k - s$. In the third case, we have $p \mid 2s$ and, since p is odd, it follows that $p \mid s$. Let $s = ps_1$. From $s^2 \equiv a \,(\text{mod } p^k)$ it follows that

$$p^2 s_1^2 = a + mp^k$$

for some m, whence $a \equiv 0 \,(\text{mod } p)$, which contradicts the hypothesis $p \nmid a$.

Exercise 3. Find all the solutions of $x^2 \equiv 44 \,(\text{mod } 125)$ by starting with $2^2 \equiv 3^2 \equiv 44 \,(\text{mod } 5)$.

Lemma 1 takes care of odd primes, but there is still the case $p = 2$. This is slightly more complicated. If we have $x^2 \equiv a \,(\text{mod } 2^e)$, first we can suppose that a is odd. (If not, we can divide out powers of 2 until we do get a congruence with odd a. For example, $x^2 \equiv 12 \,(\text{mod } 16)$ implies $(x/2)^2 \equiv 3 \,(\text{mod } 4)$.) Second, we can suppose that $a \equiv 1 \,(\text{mod } 8)$ because the square of any odd integer is congruent to 1, modulo 8. The question of solutions to $x^2 \equiv a \,(\text{mod } 2^e)$ is completely answered in

LEMMA 2. *If $a \equiv 1 \,(\text{mod } 8)$, then $x^2 \equiv a \,(\text{mod } 2^e)$ has exactly one, exactly two, or exactly four solutions, according as $e = 1$, $e = 2$, or $e \geq 3$.*

PROOF. The first two cases are obvious. So is the third for $e = 3$.

Suppose that $e \geq 4$. We will proceed in the same way as we did for odd primes to show that $x^2 \equiv a \pmod{2^e}$ has at least four solutions, by taking a solution of the congruence modulo 2^{e-1} and using it to construct a solution modulo 2^e. Suppose that $r^2 \equiv a \pmod{2^{e-1}}$. Then $r^2 = a + h2^{e-1}$ for some h. Let $R_j = r + j2^{e-2}$. Then

$$R_j = r^2 + 2rj2^{e-2} + j^2 2^{2e-4} \equiv a + (h + rj)2^{e-1} \pmod{2^e}.$$

So, to make $R_j^2 \equiv a \pmod{2^e}$, we need only choose j such that $h + rj \equiv 0 \pmod 2$. Because r is odd, this is always possible. Hence, each solution of $x^2 \equiv a \pmod{2^{e-1}}$ generates a solution of $x^2 \equiv a \pmod{2^e}$.

For an example, let us find a solution of $x^2 \equiv 17 \pmod{64}$ by starting with $1^2 \equiv 17 \pmod{16}$. Let $R_j = 1 + 8j$. Then

$$R_j^2 = 1^2 + 16j + 64j^2 \equiv 1 + 16j \pmod{32}.$$

If we want this to be $17 \pmod{32}$, we can take $j = 1$. Hence $9^2 \equiv 17 \pmod{32}$. To go up one more power, let $S_j = 9 + 16j$. Then

$$S_j^2 = 81 + 18 \cdot 16j + 256j^2 \equiv 17 + 18 \cdot 16j \pmod{64}.$$

Taking $j = 0$ yields the solution $9^2 \equiv 17 \pmod{64}$.

Exercise 4. Find a solution of $x^2 \equiv 17 \pmod{128}$.

Exercise 5. If r satisfies $x^2 \equiv a \pmod{2^e}$, show that $2^e - r$, $2^{e-1} - r$, and $2^{e-1} + r$ also satisfy it.

Exercise 6. From $9^2 \equiv 17 \pmod{64}$, find three other solutions.

It remains to show that there are no more than four solutions of $x^2 \equiv a \pmod{2^e}$ for $a \equiv 1 \pmod 8$ and $e \geq 3$. This could be done by using induction, Exercise 5, and the same ideas used in the proof of Lemma 1. Since the ideas are the same, we will omit the details.

With the aid of Lemmas 1 and 2, we can look at a congruence of the form $x^2 \equiv a \pmod m$ and say almost right away how many solutions it has. For example $x^2 \equiv 9 \pmod{1200}$ has 8 solutions, because $1200 = 2^4 \cdot 3 \cdot 5^2$, and $x^2 \equiv 9 \pmod{2^4}$ has four solutions, $x^2 \equiv 9 \pmod 3$ has one solution, and $x^2 \equiv 9 \pmod{25}$ has two solutions.

Exercise 7. How many solutions has $x^2 \equiv 10 \pmod{1200}$?

The general quadratic $Ax^2 + Bx + C \equiv 0 \pmod m$ is solved in the same way as $x^2 \equiv a \pmod m$: break it into pieces,

$$Ax^2 + Bx + C \equiv 0 \pmod{p^e},$$

start with $Ax^2 + Bx + C \equiv 0 \pmod p$, solve it (we know how from Section 11), and use the solutions to get solutions modulo $p^2, p^3, \ldots, p^e$. Then put the pieces together to get solutions modulo m. It is difficult to say in general how many solutions $Ax^2 + Bx + C \equiv 0 \pmod{m}$ has, because in going from solutions $\pmod{p^{k-1}}$ to solutions $\pmod{p^k}$, we may find one solution giving rise to p new ones, 1 new one, or no new ones, depending on the form of the quadratic. Nevertheless, all solutions of $Ax^2 + Bx + C \equiv 0 \pmod{p^e}$ will invariably be found by this method. We will not prove this, but will give two examples to show the sort of thing that can happen. Let us try to solve

(7) $$x^2 + x + 1 \equiv 0 \pmod{27}.$$

We see that $1^2 + 1 + 1 \equiv 0 \pmod 3$, so to get solutions modulo 9, let $R_j = 1 + 3j$. Then

$$R_j^2 + R_j + 1 = (1 + 6j + 9j^2) + (1 + 3j) + 1$$
$$= 3 + 9j + 9j^2 \equiv 3 \pmod 9.$$

Thus there is *no* value of j that will make $R_j^2 + R_j + 1$ congruent to 0 $\pmod 9$. Thus (7) has no solutions.

An example where one root gives rise to p others is

(8) $$x^2 + x + 7 \equiv 0 \pmod{27}.$$

$x^2 + x + 7 \equiv 0 \pmod 3$ is satisfied only for $x \equiv 1 \pmod 3$. If $R_j = 1 + 3j$, then

$$R_j^2 + R_j + 1 = 9 + 9j + 9j^2,$$

and this is congruent to zero, modulo 9, for *all j*. Hence

$$x^2 + x + 7 \equiv 0 \pmod 9$$

is satisfied by $x \equiv 1, 4,$ or $7 \pmod 9$. We must now take each of the roots up one more power to get solutions of (8). If we let $S_j = 1 + 9j$, then

$$S_j^2 + S_j + 7 = (1 + 18j + 81j^2) + (1 + 9j) + 7 \equiv 9 \pmod{27},$$

and no value of j makes the right-hand side zero, modulo 27. Thus the root 1 $\pmod 9$ generates no roots $\pmod{27}$. Next, take the root 4: if $T_j = 4 + 9j$, then

$$T_j^2 + T_j + 7 = (16 + 72j + 81j^2) + (4 + 9j) + 7$$
$$\equiv 27 + 81j \equiv 0 \pmod{27}.$$

Thus $T_j^2 + T_j + 7 \equiv 0 \pmod{27}$ for all j, and this gives three roots of (8), namely 4, 13, and 22.

Exercise 8. Show that the root 7 $\pmod 9$ gives no root $\pmod{27}$.

Thus (8) has just three solutions, $x = 4$, $x = 13$, and $x = 22$.

Problems

1. Solve

(a) $x^2 \equiv 7 \pmod{243}$,   (b) $x^2 \equiv 44 \pmod{625}$,
(c) $x^2 \equiv 17 \pmod{256}$.

2. Solve $x^2 + x + 7 \equiv 0 \pmod{81}$.

3. How many solutions have

(a) $x^2 \equiv 189 \pmod{900}$?   (b) $x^2 \equiv 89 \pmod{1000}$?
(c) $x^2 \equiv -11 \pmod{1100}$?

4. Find at least three solutions to each of the congruences in Problem 3.

5. How many solutions has $3x^2 + 2x + 1 \equiv 134 \pmod{800}$?

6. Show that any $x \equiv 33 \pmod{40}$ satisfies the congruence in the last problem.

7. How many solutions can $x^2 \equiv a \pmod{400}$ have? Construct examples illustrating each case.

8. Show that if r is a solution of $x^3 \equiv a \pmod{3^e}$ for some $e \geq 2$, then $3^{e-1} + r$ also satisfies the congruence. What about $3^{e-1} - r$ and $3^e - r$?

9. Solve (a) $x^3 \equiv 5 \pmod{16}$,   (b) $x^3 \equiv 10 \pmod{27}$.

Table A

The following table gives the smallest prime factor of each odd positive integer n, $3 \leq n \leq 9999$, not divisible by five. The numbers across the top of each column—1, 3, 7, 9—give the units' digit of n and the numbers down the side give the thousands', hundreds', and tens' digits of n. A dash in the table indicates that n is prime.

For example, reading across line 102 of the table, we see that 1021 is prime, 1023 is divisible by 3, 1027 is divisible by 13, and 1029 is divisible by 3. With the aid of the table, the prime-power decomposition of any integer less than 10,000 (and of any even integer less than 20,000) can be quickly determined. For example, take 3141. From line 314 of the table, we see that 3|3141, and a division shows that $3141 = 3 \cdot 1047$. From line 104 of the table, 3|1047, so $3141 = 3^2 \cdot 349$. Line 34 of the table shows that 349 is prime, so we have the prime-power decomposition of 3141.

	1	3	7	9		1	3	7	9		1	3	7	9		1	3	7	9
0	—	—		3	**20**	3	7	3	11	**40**	—	13	11	—	**60**	—	3	—	3
1	—	—	—	—	21	—	3	7	3	41	3	7	3	—	61	13	—	—	—
2	3	—	3	—	22	13	—	—	—	42	—	3	7	3	62	3	7	3	17
3	—	3	—	3	23	3	—	3	—	43	—	—	19	—	63	—	3	7	3
4	—	—	—	7	24	—	3	13	3	44	3	—	3	—	64	—	—	—	11
5	3	—	3	—	25	—	11	—	7	45	11	3	—	3	65	3	—	3	—
6	—	3	—	3	26	3	—	3	—	46	—	—	—	7	66	—	3	23	3
7	—	—	7	—	27	—	3	—	3	47	3	11	3	—	67	11	—	—	7
8	3	—	3	—	28	—	—	7	17	48	13	3	—	3	68	3	—	3	13
9	7	3	—	3	29	3	—	3	13	49	—	17	7	—	69	—	3	17	3
10	—	—	—	—	**30**	7	3	—	3	**50**	3	—	3	—	**70**	—	19	7	—
11	3	—	3	7	31	—	—	—	11	51	7	3	11	3	71	3	23	3	—
12	11	3	—	3	32	3	17	3	7	52	—	—	17	23	72	7	3	—	3
13	—	7	—	—	33	—	3	—	3	53	3	13	3	7	73	17	—	11	—
14	3	11	3	—	34	11	7	—	—	54	—	3	—	3	74	3	—	3	7
15	—	3	—	3	35	3	—	3	—	55	19	7	—	13	75	—	3	—	3
16	7	—	—	13	36	19	3	—	3	56	3	—	3	—	76	—	7	13	—
17	3	—	3	—	37	7	—	13	—	57	—	3	—	3	77	3	—	3	19
18	—	3	11	3	38	3	—	3	—	58	7	11	—	19	78	11	3	—	3
19	—	—	—	—	39	17	3	—	3	59	3	—	3	—	79	7	13	—	17

Table A Continued

	1	3	7	9		1	3	7	9		1	3	7	9		1	3	7	9
80	3	11	3	—	**120**	—	3	17	3	**160**	—	7	—	—	**200**	3	—	3	7
81	—	3	19	3	121	7	—	—	23	161	3	—	3	—	201	—	3	—	3
82	—	—	—	—	122	3	—	3	—	162	—	3	—	3	202	43	7	—	—
83	3	7	3	—	123	—	3	—	3	163	7	23	—	11	203	3	19	3	—
84	29	3	7	3	124	17	11	29	—	164	3	31	3	17	204	13	3	23	3
85	23	—	—	—	125	3	7	3	—	165	13	3	—	3	205	7	—	11	29
86	3	—	3	11	126	13	3	7	3	166	11	—	—	—	206	3	—	3	—
87	13	3	—	3	127	31	19	—	—	167	3	7	3	23	207	19	3	31	3
88	—	—	—	7	128	3	—	3	—	168	41	3	7	3	208	—	—	—	—
89	3	19	3	29	129	—	3	—	3	169	19	—	—	—	209	3	7	3	—
90	17	3	—	3	**130**	—	—	—	7	**170**	3	13	3	—	**210**	11	3	7	3
91	—	11	7	—	131	3	13	3	—	171	29	3	17	3	211	—	—	29	13
92	3	13	3	—	132	—	3	—	3	172	—	—	11	7	212	3	11	3	—
93	7	3	—	3	133	11	31	7	13	173	3	—	3	37	213	—	3	—	3
94	—	23	—	13	134	3	17	3	19	174	—	3	—	3	214	—	—	19	7
95	3	—	3	7	135	7	3	23	3	175	17	—	7	—	215	3	—	3	17
96	31	3	—	3	136	—	29	—	37	176	3	41	3	29	216	—	3	11	3
97	—	7	—	11	137	3	—	3	7	177	7	3	—	3	217	13	41	7	—
98	3	—	3	23	138	—	3	19	3	178	13	—	—	—	218	3	37	3	11
99	—	3	—	3	139	13	7	11	—	179	3	11	3	7	219	7	3	13	3
100	7	17	19	—	**140**	3	23	3	—	**180**	—	3	13	3	**220**	31	—	—	47
101	3	—	3	—	141	17	3	13	3	181	—	7	23	17	221	3	—	3	7
102	—	3	13	3	142	7	—	—	—	182	3	—	3	31	222	—	3	17	3
103	—	—	17	—	143	3	—	3	—	183	—	3	11	3	223	23	7	—	—
104	3	7	3	—	144	11	3	—	3	184	7	19	—	43	224	3	—	3	13
105	—	3	7	3	145	—	—	31	—	185	3	17	3	11	225	—	3	37	3
106	—	—	11	—	146	3	7	3	13	186	—	3	—	3	226	7	31	—	—
107	3	29	3	13	147	—	3	7	3	187	—	—	—	—	227	3	—	3	43
108	23	3	—	3	148	—	—	—	—	188	3	7	3	—	228	—	3	—	3
109	—	—	—	7	149	3	—	3	—	189	31	3	7	3	229	29	—	—	11
110	3	—	3	—	**150**	19	3	11	3	**190**	—	11	—	23	**230**	3	7	3	—
111	11	3	—	3	151	—	17	37	7	191	3	—	3	19	231	—	3	7	3
112	19	—	7	—	152	3	—	3	11	192	17	3	41	3	232	11	23	13	17
113	3	11	3	17	153	—	3	29	3	193	—	—	13	7	233	3	—	3	—
114	7	3	31	3	154	23	—	7	—	194	3	29	3	—	234	—	3	—	3
115	—	—	13	19	155	3	—	3	—	195	—	3	19	3	235	—	13	—	7
116	3	—	3	7	156	7	3	—	3	196	37	13	7	11	236	3	17	3	23
117	—	3	11	3	157	—	11	19	—	197	3	—	3	—	237	—	3	—	3
118	—	7	—	29	158	3	—	3	7	198	7	3	—	3	238	—	—	7	—
119	3	—	3	11	159	37	3	—	3	199	11	—	—	—	239	3	—	3	—

Table A

	1	3	7	9		1	3	7	9		1	3	7	9		1	3	7	9
240	7	3	29	3	**280**	—	—	7	53	**320**	3	—	3	—	**360**	13	3	—	3
241	—	19	—	41	281	3	29	3	—	321	13	3	—	3	361	23	—	—	7
242	3	—	3	7	282	7	3	11	3	322	—	11	7	—	362	3	—	3	19
243	11	3	—	3	283	19	—	—	17	323	3	53	3	41	363	—	3	—	3
244	—	7	—	31	284	3	—	3	7	324	7	3	17	3	364	11	—	7	41
245	3	11	3	—	285	—	3	—	3	325	—	—	—	—	365	3	13	3	—
246	23	3	—	3	286	—	7	47	19	326	3	13	3	7	366	7	3	19	3
247	7	—	—	37	287	3	13	3	—	327	—	3	29	3	367	—	—	—	13
248	3	13	3	19	288	43	3	—	3	328	17	7	19	11	368	3	29	3	7
249	47	3	11	3	289	7	11	—	13	329	3	37	3	—	369	—	3	—	3
250	41	—	23	13	**290**	3	—	3	—	**330**	—	3	—	3	**370**	—	7	11	—
251	3	7	3	11	291	41	3	—	3	331	7	—	31	—	371	3	47	3	—
252	—	3	7	3	292	23	37	—	29	332	3	—	3	—	372	61	3	—	3
253	—	17	43	—	293	3	7	3	—	333	—	3	47	3	373	7	—	37	—
254	3	—	3	—	294	17	3	7	3	334	13	—	—	17	374	3	19	3	23
255	—	3	—	3	295	13	—	—	11	335	3	7	3	—	375	11	3	13	3
256	13	11	17	7	296	3	—	3	—	336	—	3	7	3	376	—	53	—	—
257	3	31	3	—	297	—	3	13	3	337	—	—	11	31	377	3	7	3	—
258	29	3	13	3	298	11	19	29	7	338	3	17	3	—	378	19	3	7	3
259	—	—	7	23	299	3	41	3	—	339	—	3	43	3	379	17	—	—	29
260	3	19	3	—	**300**	—	3	31	3	**340**	19	41	—	7	**380**	3	—	3	13
261	7	3	—	3	301	—	23	7	—	341	3	—	3	13	381	37	3	11	3
262	—	43	37	11	302	3	—	3	13	342	11	3	23	3	382	—	—	43	7
263	3	—	3	7	303	7	3	—	3	343	47	—	7	19	383	3	—	3	11
264	19	3	—	3	304	—	17	11	—	344	3	11	3	—	384	23	3	—	3
265	11	7	—	—	305	3	43	3	7	345	7	3	—	3	385	—	—	7	17
266	3	—	3	17	306	—	3	—	3	346	—	—	—	—	386	3	—	3	53
267	—	3	—	3	307	37	7	17	—	347	3	23	3	7	387	7	3	—	3
268	7	—	—	—	308	3	—	3	—	348	59	3	11	3	388	—	11	13	—
269	3	—	3	—	309	11	3	19	3	349	—	7	13	—	389	3	17	3	7
270	37	3	—	3	**310**	7	29	13	—	**350**	3	31	3	11	**390**	47	3	—	3
271	—	—	11	—	311	3	11	3	—	351	—	3	—	3	391	—	7	—	—
272	3	7	3	—	312	—	3	53	3	352	7	13	—	—	392	3	—	3	—
273	—	3	7	3	313	31	13	—	43	353	3	—	3	—	393	—	3	31	3
274	—	13	41	—	314	3	7	3	47	354	—	3	—	3	394	7	—	—	11
275	3	—	3	31	315	23	3	7	3	355	53	11	—	—	395	3	59	3	37
276	11	3	—	3	316	29	—	—	—	356	3	7	3	43	396	17	3	—	3
277	17	47	—	7	317	3	19	3	11	357	—	3	7	3	397	11	29	41	23
278	3	11	3	—	318	—	3	—	3	358	—	—	17	37	398	3	7	3	—
279	—	3	—	3	319	—	31	23	7	359	3	—	3	59	399	13	3	7	3

Table A Continued

	1	3	7	9		1	3	7	9		1	3	7	9		1	3	7	9
400	—	—	—	19	**440**	3	7	3	—	**480**	—	3	11	3	**520**	7	11	41	—
401	3	—	3	—	441	11	3	7	3	481	17	—	—	61	521	3	13	3	17
402	—	3	—	3	442	—	—	19	43	482	3	7	3	11	522	23	3	—	3
403	29	37	11	7	443	3	11	3	23	483	—	3	7	3	523	—	—	—	13
404	3	13	3	—	444	—	3	—	3	484	47	29	37	13	524	3	7	3	29
405	—	3	—	3	445	—	61	—	7	485	3	23	3	43	525	59	3	7	3
406	31	17	7	13	446	3	—	3	41	486	—	3	31	3	526	—	19	23	11
407	3	—	3	—	447	17	3	11	3	487	—	11	—	7	527	3	—	3	—
408	7	3	61	3	448	—	—	7	67	488	3	19	3	—	528	—	3	17	3
409	—	—	17	—	449	3	—	3	11	489	67	3	59	3	529	11	67	—	7
410	3	11	3	7	**450**	7	3	—	3	**490**	13	—	7	—	**530**	3	—	3	—
411	—	3	23	3	451	13	—	—	—	491	3	17	3	—	531	47	3	13	3
412	13	7	—	—	452	3	—	3	7	492	7	3	13	3	532	17	—	7	73
413	3	—	3	—	453	23	3	13	3	493	—	—	—	11	533	3	—	3	19
414	41	3	11	3	454	19	7	—	—	494	3	—	3	7	534	7	3	—	3
415	7	—	—	—	455	3	29	3	47	495	—	3	—	3	535	—	53	11	23
416	3	23	3	11	456	—	3	—	3	496	11	7	—	—	536	3	31	3	7
417	43	3	—	3	457	7	17	23	19	497	3	—	3	13	537	41	3	19	3
418	37	47	53	59	458	3	—	3	13	498	17	3	—	3	538	—	7	—	17
419	3	7	3	13	459	—	3	—	3	499	7	—	19	—	539	3	—	3	—
420	—	3	7	3	**460**	43	—	17	11	**500**	3	—	3	—	**540**	11	3	—	3
421	—	11	—	—	461	3	7	3	31	501	—	3	29	3	541	7	—	—	—
422	3	41	3	—	462	—	3	7	3	502	—	—	11	47	542	3	11	3	61
423	—	3	19	3	463	11	41	—	—	503	3	7	3	—	543	—	3	—	3
424	—	—	31	7	464	3	—	3	—	504	71	3	7	3	544	—	—	13	—
425	3	—	3	—	465	—	3	—	3	505	—	31	13	—	545	3	7	3	53
426	—	3	17	3	466	59	—	13	7	506	3	61	3	37	546	43	3	7	3
427	—	—	7	11	467	3	—	3	—	507	11	3	—	3	547	—	13	—	—
428	3	—	3	—	468	31	3	43	3	508	—	13	—	7	548	3	—	3	11
429	7	3	—	3	469	—	13	7	37	509	3	11	3	—	549	17	3	23	3
430	11	13	59	31	**470**	3	—	3	17	**510**	—	3	—	3	**550**	—	—	—	7
431	3	19	3	7	471	7	3	53	3	511	19	—	7	—	551	3	37	3	—
432	29	3	—	3	472	—	—	29	—	512	3	47	3	23	552	—	3	—	3
433	61	7	—	—	473	3	—	3	7	513	7	3	11	3	553	—	11	7	29
434	3	43	3	—	474	11	3	47	3	514	53	37	—	19	554	3	23	3	31
435	19	3	—	3	475	—	7	67	—	515	3	—	3	7	555	7	3	—	3
436	7	—	11	17	476	3	11	3	19	516	13	3	—	3	556	67	—	19	—
437	3	—	3	29	477	13	3	17	3	517	—	7	31	—	557	3	—	3	7
438	13	3	41	3	478	7	—	—	—	518	3	71	3	—	558	—	3	37	3
439	—	23	—	53	479	3	—	3	—	519	29	3	—	3	559	—	7	29	11

Table A

	1	3	7	9		1	3	7	9		1	3	7	9		1	3	7	9
560	3	13	3	71	**600**	17	3	—	3	**640**	37	19	43	13	**680**	3	—	3	11
561	31	3	41	3	601	—	7	11	13	641	3	11	3	7	681	7	3	17	3
562	7	—	17	13	602	3	19	3	—	642	—	3	—	3	682	19	—	—	—
563	3	43	3	—	603	37	3	—	3	643	59	7	41	47	683	3	—	3	7
564	—	3	—	3	604	7	—	—	23	644	3	17	3	—	684	—	3	41	3
565	—	—	—	—	605	3	—	3	73	645	—	3	11	3	685	13	7	—	19
566	3	7	3	—	606	11	3	—	3	646	7	23	29	—	686	3	—	3	—
567	53	3	7	3	607	13	—	59	—	647	3	—	3	11	687	—	3	13	3
568	13	—	11	—	608	3	7	3	—	648	—	3	13	3	688	7	—	71	83
569	3	—	3	41	609	—	3	7	3	649	—	43	73	67	689	3	61	3	—
570	—	3	13	3	**610**	—	17	31	41	**650**	3	7	3	23	**690**	67	3	—	3
571	—	29	—	7	611	3	—	3	29	651	17	3	7	3	691	—	31	—	11
572	3	59	3	17	612	—	3	11	3	652	—	11	61	—	692	3	7	3	13
573	11	3	—	3	613	—	—	17	7	653	3	47	3	13	693	29	3	7	3
574	—	—	7	—	614	3	—	3	11	654	31	3	—	3	694	11	53	—	—
575	3	11	3	13	615	—	3	47	3	655	—	—	79	7	695	3	17	3	—
576	7	3	73	3	616	61	—	7	31	656	3	—	3	—	696	—	3	—	3
577	29	23	53	—	617	3	—	3	37	657	—	3	—	3	697	—	19	—	7
578	3	—	3	7	618	7	3	23	3	658	—	29	7	11	698	3	—	3	29
579	—	3	11	3	619	41	11	—	—	659	3	19	3	—	699	—	3	—	3
580	—	7	—	37	**620**	3	—	3	7	**660**	7	3	—	3	**700**	—	47	7	43
581	3	—	3	11	621	—	3	—	3	661	11	17	13	—	701	3	—	3	—
582	—	3	—	3	622	—	7	13	—	662	3	37	3	7	702	7	3	—	3
583	7	19	13	—	623	3	23	3	17	663	19	3	—	3	703	79	13	31	—
584	3	—	3	—	624	79	3	—	3	664	29	7	17	61	704	3	—	3	7
585	—	3	—	3	625	7	13	—	11	665	3	—	3	—	705	11	3	—	3
586	—	11	—	—	626	3	—	3	—	666	—	3	59	3	706	23	7	37	—
587	3	7	3	—	627	—	3	—	3	667	7	—	11	—	707	3	11	3	—
588	—	3	7	3	628	11	61	—	19	668	3	41	3	—	708	73	3	19	3
589	43	71	—	17	629	3	7	3	—	669	—	3	37	3	709	7	41	47	31
590	3	—	3	19	**630**	—	3	7	3	**670**	—	—	19	—	**710**	3	—	3	—
591	23	3	61	3	631	—	59	—	71	671	3	7	3	—	711	13	3	11	3
592	31	—	—	7	632	3	—	3	—	672	11	3	7	3	712	—	17	—	—
593	3	17	3	—	633	13	3	—	3	673	53	—	—	23	713	3	7	3	11
594	13	3	19	3	634	17	—	11	7	674	3	11	3	17	714	37	3	7	3
595	11	—	7	59	635	3	—	3	—	675	43	3	29	3	715	—	23	17	—
596	3	67	3	47	636	—	3	—	3	676	—	—	67	7	716	3	13	3	67
597	7	3	43	3	637	23	—	7	—	677	3	13	3	—	717	71	3	—	3
598	—	31	—	53	638	3	13	3	—	678	—	3	11	3	718	43	11	—	7
599	3	13	3	7	639	7	3	—	3	679	—	—	7	13	719	3	—	3	23

Table A Continued

	1	3	7	9		1	3	7	9		1	3	7	9		1	3	7	9
720	19	3	—	3	**760**	11	—	—	7	**800**	3	53	3	—	**840**	31	3	7	3
721	—	—	7	—	761	3	23	3	19	801	—	3	—	3	841	13	47	19	—
722	3	31	3	—	762	—	3	29	3	802	13	71	23	7	842	3	—	3	—
723	7	3	—	3	763	13	17	7	—	803	3	29	3	—	843	—	3	11	3
724	13	—	—	11	764	3	—	3	—	804	11	3	13	3	844	23	—	—	7
725	3	—	3	7	765	7	3	13	3	805	83	—	7	—	845	3	79	3	11
726	53	3	13	3	766	47	79	11	—	806	3	11	3	—	846	—	3	—	3
727	11	7	19	29	767	3	—	3	7	807	7	3	41	3	847	43	37	7	61
728	3	—	3	37	768	—	3	—	3	808	—	59	—	—	848	3	17	3	13
729	23	3	—	3	769	—	7	43	—	809	3	—	3	7	849	7	3	29	3
730	7	67	—	—	**770**	3	—	3	13	**810**	—	3	11	3	**850**	—	11	47	67
731	3	71	3	13	771	11	3	—	3	811	—	7	—	23	851	3	—	3	7
732	—	3	17	3	772	7	—	—	59	812	3	—	3	11	852	—	3	—	3
733	—	—	11	41	773	3	11	3	71	813	47	3	79	3	853	19	7	—	—
734	3	7	3	—	774	—	3	61	3	814	7	17	—	29	854	3	—	3	83
735	—	3	7	3	775	23	—	—	—	815	3	31	3	41	855	17	3	43	3
736	17	37	53	—	776	3	7	3	17	816	—	3	—	3	856	7	—	13	11
737	3	73	3	47	777	19	3	7	3	817	—	11	13	—	857	3	—	3	23
738	11	3	83	3	778	31	43	13	—	818	3	7	3	19	858	—	3	31	3
739	19	—	13	7	779	3	—	3	11	819	—	3	7	3	859	11	13	—	—
740	3	11	3	31	**780**	29	3	37	3	**820**	59	13	29	—	**860**	3	7	3	—
741	—	3	—	3	781	73	13	—	7	821	3	43	3	—	861	79	3	7	3
742	41	13	7	17	782	3	—	3	—	822	—	3	19	3	862	37	—	—	—
743	3	—	3	43	783	41	3	17	3	823	—	—	—	7	863	3	89	3	53
744	7	3	11	3	784	—	11	7	47	824	3	—	3	73	864	—	3	—	3
745	—	29	—	—	785	3	—	3	29	825	37	3	23	3	865	41	17	11	7
746	3	17	3	7	786	7	3	—	3	826	11	—	7	—	866	3	—	3	—
747	31	3	—	3	787	17	—	—	—	827	3	—	3	17	867	13	3	—	3
748	—	7	—	—	788	3	—	3	7	828	7	3	—	3	868	—	19	7	—
749	3	59	3	—	789	13	3	53	3	829	—	—	—	43	869	3	—	3	—
750	13	3	—	3	**790**	—	7	—	11	**830**	3	19	3	7	**870**	7	3	—	3
751	7	11	—	73	791	3	41	3	—	831	—	3	—	3	871	31	—	23	—
752	3	—	3	—	792	89	3	—	3	832	53	7	11	—	872	3	11	3	7
753	17	3	—	3	793	7	—	—	17	833	3	13	3	31	873	—	3	—	3
754	—	19	—	—	794	3	13	3	—	834	19	3	17	3	874	—	7	—	13
755	3	7	3	—	795	—	3	73	3	835	7	—	61	13	875	3	—	3	19
756	—	3	7	3	796	19	—	31	13	836	3	—	3	—	876	—	3	11	3
757	67	—	—	11	797	3	7	3	79	837	11	3	—	3	877	7	31	67	—
758	3	—	3	—	798	23	3	7	3	838	17	83	—	—	878	3	—	3	11
759	—	3	71	3	799	61	—	11	19	839	3	7	3	37	879	59	3	19	3

Table A

	1	3	7	9		1	3	7	9		1	3	7	9		1	3	7	9
880	13	—	—	23	**910**	19	—	7	—	**940**	7	—	23	97	**970**	89	31	17	7
881	3	7	3	—	911	3	13	3	11	941	3	—	3	—	971	3	11	3	—
882	—	3	7	3	912	7	3	—	3	942	—	3	11	3	972	—	3	71	3
883	—	11	—	—	913	23	—	—	13	943	—	—	—	—	973	37	—	7	—
884	3	37	3	—	914	3	41	3	7	944	3	7	3	11	974	3	—	3	—
885	53	3	17	3	915	—	3	—	3	945	13	3	7	3	975	7	3	11	3
886	—	—	—	7	916	—	7	89	53	946	—	—	—	17	976	43	13	—	—
887	3	19	3	13	917	3	—	3	67	947	3	—	3	—	977	3	29	3	7
888	83	3	—	3	918	—	3	—	3	948	19	3	53	3	978	—	3	—	3
889	17	—	7	11	919	7	29	17	—	949	—	11	—	7	979	—	7	97	41
890	3	29	3	59	**920**	3	—	3	—	**950**	3	13	3	37	**980**	3	—	3	17
891	7	3	37	3	921	61	3	13	3	951	—	3	31	3	981	—	3	—	3
892	11	—	79	—	922	—	23	—	11	952	—	89	7	13	982	7	11	31	—
893	3	—	3	7	923	3	7	3	—	953	3	—	3	—	983	3	—	3	—
894	—	3	23	3	924	—	3	7	3	954	7	3	—	3	984	13	3	43	3
895	—	7	13	17	925	11	19	—	47	955	—	41	19	11	985	—	59	—	—
896	3	—	3	—	926	3	59	3	13	956	3	73	3	7	986	3	7	3	71
897	—	3	47	3	927	73	3	—	3	957	17	3	61	3	987	—	3	7	3
898	7	13	11	89	928	—	—	37	7	958	11	7	—	43	988	41	—	—	11
899	3	17	3	—	929	3	—	3	17	959	3	53	3	29	989	3	13	3	19
900	—	3	—	3	**930**	71	3	41	3	**960**	—	3	13	3	**990**	—	3	—	3
901	—	—	71	29	931	—	67	7	—	961	7	—	59	—	991	11	23	47	7
902	3	7	3	—	932	3	—	3	19	962	3	—	3	—	992	3	—	3	—
903	11	3	7	3	933	7	3	—	3	963	—	3	23	3	993	—	3	19	3
904	—	—	83	—	934	—	—	13	—	964	31	—	11	—	994	—	61	7	—
905	3	11	3	—	935	3	47	3	7	965	3	7	3	13	995	3	37	3	23
906	13	3	—	3	936	11	3	17	3	966	—	3	7	3	996	7	3	—	3
907	47	43	29	7	937	—	7	—	83	967	19	17	—	—	997	13	—	11	17
908	3	31	3	61	938	3	11	3	41	968	3	23	3	—	998	3	67	3	7
909	—	3	11	3	939	—	3	—	3	969	11	3	—	3	999	97	3	13	3

Table B

The following table lists all the perfect squares less than 200,000. These are the squares of integers n, $0 \le n \le 447$. To locate n^2 in the table, look at the intersection of the column headed by the units' digit of n with line $[n/10]$. For example, the square of 314 is located on line 31, in column 4: $314^2 = 98{,}596$.

	0	1	2	3	4	5	6	7	8	9
0	0	1	4	9	16	25	36	49	64	81
1	100	121	144	169	196	225	256	289	324	361
2	400	441	484	529	576	625	676	729	784	841
3	900	961	1024	1089	1156	1225	1296	1369	1444	1521
4	1600	1681	1764	1849	1936	2025	2116	2209	2304	2401
5	2500	2601	2704	2809	2916	3025	3136	3249	3364	3481
6	3600	3721	3844	3969	4096	4225	4356	4489	4624	4761
7	4900	5041	5184	5329	5476	5625	5776	5929	6084	6241
8	6400	6561	6724	6889	7056	7225	7396	7569	7744	7921
9	8100	8281	8464	8649	8836	9025	9216	9409	9604	9801
10	10000	10201	10404	10609	10816	11025	11236	11449	11664	11881
11	12100	12321	12544	12769	12996	13225	13456	13689	13924	14161
12	14400	14641	14884	15129	15376	15625	15876	16129	16384	16641
13	16900	17161	17424	17689	17956	18225	18496	18769	19044	19321
14	19600	19881	20164	20449	20736	21025	21316	21609	21904	22201
15	22500	22801	23104	23409	23716	24025	24336	24649	24964	25281
16	25600	25921	26244	26569	26896	27225	27556	27889	28224	28561
17	28900	29241	29584	29929	30276	30625	30976	31329	31684	32041
18	32400	32761	33124	33489	33856	34225	34596	34969	35344	35721
19	36100	36481	36864	37249	37636	38025	38416	38809	39204	39601

	0	1	2	3	4	5	6	7	8	9
20	40000	40401	40804	41209	41616	42025	42436	42849	43264	43681
21	44100	44521	44944	45369	45796	46225	46656	47089	47524	47961
22	48400	48841	49284	49729	50176	50625	51076	51529	51984	52441
23	52900	53361	53824	54289	54756	55225	55696	56169	56644	57121
24	57600	58081	58564	59049	59536	60025	60516	61009	61504	62001
25	62500	63001	63504	64009	64516	65025	65536	66049	66564	67081
26	67600	68121	68644	69169	69696	70225	70756	71289	71824	72361
27	72900	73441	73984	74529	75076	75625	76176	76729	77284	77841
28	78400	78961	79524	80089	80656	81225	81796	82369	82944	83521
29	84100	84681	85264	85849	86436	87025	87616	88209	88804	89401
30	90000	90601	91204	91809	92416	93025	93636	94249	94864	95481
31	96100	96721	97344	97969	98596	99225	99856	100489	101124	101761
32	102400	103041	103684	104329	104976	105625	106276	106929	107584	108241
33	108900	109561	110224	110889	111556	112225	112896	113569	114244	114921
34	115600	116281	116964	117649	118336	119025	119716	120409	121104	121801
35	122500	123201	123904	124609	125316	126025	126736	127449	128164	128881
36	129600	130321	131044	131769	132496	133225	133956	134689	135424	136161
37	136900	137641	138384	139129	139876	140625	141376	142129	142884	143641
38	144400	145161	145924	146689	147456	148225	148996	149769	150544	151321
39	152100	152881	153664	154449	155236	156025	156816	157609	158404	159201
40	160000	160801	161604	162409	163216	164025	164836	165649	166464	167281
41	168100	168921	169744	170569	171396	172225	173056	173889	174724	175561
42	176400	177241	178084	178929	179776	180625	181476	182329	183184	184041
43	184900	185761	186624	187489	188356	189225	190096	190969	191844	192721
44	193600	194481	195364	196249	197136	198025	198916	199809		

Table C

The following table gives the complete prime-power decomposition of integers n, $10,000 \leq n \leq 10,269$ and $100,000 \leq n \leq 100,149$.

10000	$2^4 \cdot 5^4$	**10030**	$2 \cdot 5 \cdot 17 \cdot 59$	**10060**	$2^2 \cdot 5 \cdot 503$
10001	$73 \cdot 137$	10031	$7 \cdot 1433$	10061	10061
10002	$2 \cdot 3 \cdot 1667$	10032	$2^4 \cdot 3 \cdot 11 \cdot 19$	10062	$2 \cdot 3^2 \cdot 13 \cdot 43$
10003	$7 \cdot 1429$	10033	$79 \cdot 127$	10063	$29 \cdot 347$
10004	$2^2 \cdot 41 \cdot 61$	10034	$2 \cdot 29 \cdot 173$	10064	$2^4 \cdot 17 \cdot 37$
10005	$3 \cdot 5 \cdot 23 \cdot 29$	10035	$3^2 \cdot 5 \cdot 223$	10065	$3 \cdot 5 \cdot 11 \cdot 61$
10006	$2 \cdot 5003$	10036	$2^2 \cdot 13 \cdot 193$	10066	$2 \cdot 7 \cdot 719$
10007	10007	10037	10037	10067	10067
10008	$2^3 \cdot 3^2 \cdot 139$	10038	$2 \cdot 3 \cdot 7 \cdot 239$	10068	$2^2 \cdot 3 \cdot 839$
10009	10009	10039	10039	10069	10069
10010	$2 \cdot 5 \cdot 7 \cdot 11 \cdot 13$	**10040**	$2^3 \cdot 5 \cdot 251$	**10070**	$2 \cdot 5 \cdot 19 \cdot 53$
10011	$3 \cdot 47 \cdot 71$	10041	$3 \cdot 3347$	10071	$3^3 \cdot 373$
10012	$2^2 \cdot 2503$	10042	$2 \cdot 5021$	10072	$2^3 \cdot 1259$
10013	$17 \cdot 19 \cdot 31$	10043	$11^2 \cdot 83$	10073	$7 \cdot 1439$
10014	$2 \cdot 3 \cdot 1669$	10044	$2^2 \cdot 3^4 \cdot 31$	10074	$2 \cdot 3 \cdot 23 \cdot 73$
10015	$5 \cdot 2003$	10045	$5 \cdot 7^2 \cdot 41$	10075	$5^2 \cdot 13 \cdot 31$
10016	$2^5 \cdot 313$	10046	$2 \cdot 5023$	10076	$2^2 \cdot 11 \cdot 229$
10017	$3^3 \cdot 7 \cdot 53$	10047	$3 \cdot 17 \cdot 197$	10077	$3 \cdot 3359$
10018	$2 \cdot 5009$	10048	$2^6 \cdot 157$	10078	$2 \cdot 5039$
10019	$43 \cdot 233$	10049	$13 \cdot 773$	10079	10079
10020	$2^2 \cdot 3 \cdot 5 \cdot 167$	**10050**	$2 \cdot 3 \cdot 5^2 \cdot 67$	**10080**	$2^5 \cdot 3^2 \cdot 5 \cdot 7$
10021	$11 \cdot 911$	10051	$19 \cdot 23^2$	10081	$17 \cdot 593$
10022	$2 \cdot 5011$	10052	$2^2 \cdot 7 \cdot 359$	10082	$2 \cdot 71^2$
10023	$3 \cdot 13 \cdot 257$	10053	$3^2 \cdot 1117$	10083	$3 \cdot 3361$
10024	$2^3 \cdot 7 \cdot 179$	10054	$2 \cdot 11 \cdot 457$	10084	$2^2 \cdot 2521$
10025	$5^2 \cdot 401$	10055	$5 \cdot 2011$	10085	$5 \cdot 2017$
10026	$2 \cdot 3^2 \cdot 557$	10056	$2^3 \cdot 3 \cdot 419$	10086	$2 \cdot 3 \cdot 41^2$
10027	$37 \cdot 271$	10057	$89 \cdot 113$	10087	$7 \cdot 11 \cdot 131$
10028	$2^2 \cdot 23 \cdot 109$	10058	$2 \cdot 47 \cdot 107$	10088	$2^3 \cdot 13 \cdot 97$
10029	$3 \cdot 3343$	10059	$3 \cdot 7 \cdot 479$	10089	$3^2 \cdot 19 \cdot 59$

Table C

10090	$2 \cdot 5 \cdot 1009$	**10130**	$2 \cdot 5 \cdot 1013$	**10170**	$2 \cdot 3^2 \cdot 5 \cdot 113$
10091	10091	10131	$3 \cdot 11 \cdot 307$	10171	$7 \cdot 1453$
10092	$2^2 \cdot 3 \cdot 29^2$	10132	$2^2 \cdot 17 \cdot 149$	10172	$2^2 \cdot 2543$
10093	10093	10133	10133	10173	$3 \cdot 3391$
10094	$2 \cdot 7^2 \cdot 103$	10134	$2 \cdot 3^2 \cdot 563$	10174	$2 \cdot 5087$
10095	$3 \cdot 5 \cdot 673$	10135	$5 \cdot 2027$	10175	$5^2 \cdot 11 \cdot 37$
10096	$2^4 \cdot 631$	10136	$2^3 \cdot 7 \cdot 181$	10176	$2^6 \cdot 3 \cdot 53$
10097	$23 \cdot 439$	10137	$3 \cdot 31 \cdot 109$	10177	10177
10098	$2 \cdot 3^3 \cdot 11 \cdot 17$	10138	$2 \cdot 37 \cdot 137$	10178	$2 \cdot 7 \cdot 727$
10099	10099	10139	10139	10179	$3^3 \cdot 13 \cdot 29$
10100	$2^2 \cdot 5^2 \cdot 101$	**10140**	$2^2 \cdot 3 \cdot 5 \cdot 13^2$	**10180**	$2^2 \cdot 5 \cdot 509$
10101	$3 \cdot 7 \cdot 13 \cdot 37$	10141	10141	10181	10181
10102	$2 \cdot 5051$	10142	$2 \cdot 11 \cdot 461$	10182	$2 \cdot 3 \cdot 1697$
10103	10103	10143	$3^2 \cdot 7^2 \cdot 23$	10183	$17 \cdot 599$
10104	$2^3 \cdot 3 \cdot 421$	10144	$2^5 \cdot 317$	10184	$2^3 \cdot 19 \cdot 67$
10105	$5 \cdot 43 \cdot 47$	10145	$5 \cdot 2029$	10185	$3 \cdot 5 \cdot 7 \cdot 97$
10106	$2 \cdot 31 \cdot 163$	10146	$2 \cdot 3 \cdot 19 \cdot 89$	10186	$2 \cdot 11 \cdot 463$
10107	$3^2 \cdot 1123$	10147	$73 \cdot 139$	10187	$61 \cdot 167$
10108	$2^2 \cdot 7 \cdot 19^2$	10148	$2^2 \cdot 43 \cdot 59$	10188	$2^2 \cdot 3^2 \cdot 283$
10109	$11 \cdot 919$	10149	$3 \cdot 17 \cdot 199$	10189	$23 \cdot 443$
10110	$2 \cdot 3 \cdot 5 \cdot 337$	**10150**	$2 \cdot 5^2 \cdot 7 \cdot 29$	**10190**	$2 \cdot 5 \cdot 1019$
10111	10111	10151	10151	10191	$3 \cdot 43 \cdot 79$
10112	$2^7 \cdot 79$	10152	$2^3 \cdot 3^3 \cdot 47$	10192	$2^4 \cdot 7^2 \cdot 13$
10113	$3 \cdot 3371$	10153	$11 \cdot 13 \cdot 71$	10193	10193
10114	$2 \cdot 13 \cdot 389$	10154	$2 \cdot 5077$	10194	$2 \cdot 3 \cdot 1699$
10115	$5 \cdot 7 \cdot 17^2$	10155	$3 \cdot 5 \cdot 677$	10195	$5 \cdot 2039$
10116	$2^2 \cdot 3^2 \cdot 281$	10156	$2^2 \cdot 2539$	10196	$2^2 \cdot 2549$
10117	$67 \cdot 151$	10157	$7 \cdot 1451$	10197	$3^2 \cdot 11 \cdot 103$
10118	$2 \cdot 5059$	10158	$2 \cdot 3 \cdot 1693$	10198	$2 \cdot 5099$
10119	$3 \cdot 3373$	10159	10159	10199	$7 \cdot 31 \cdot 47$
10120	$2^3 \cdot 5 \cdot 11 \cdot 23$	**10160**	$2^4 \cdot 5 \cdot 127$	**10200**	$2^3 \cdot 3 \cdot 5^2 \cdot 17$
10121	$29 \cdot 349$	10161	$3^2 \cdot 1129$	10201	101^2
10122	$2 \cdot 3 \cdot 7 \cdot 241$	10162	$2 \cdot 5081$	10202	$2 \cdot 5101$
10123	$53 \cdot 191$	10163	10163	10203	$3 \cdot 19 \cdot 179$
10124	$2^2 \cdot 2531$	10164	$2^2 \cdot 3 \cdot 7 \cdot 11^2$	10204	$2^2 \cdot 2551$
10125	$3^4 \cdot 5^3$	10165	$5 \cdot 19 \cdot 107$	10205	$5 \cdot 13 \cdot 157$
10126	$2 \cdot 61 \cdot 83$	10166	$2 \cdot 13 \cdot 17 \cdot 23$	10206	$2 \cdot 3^6 \cdot 7$
10127	$13 \cdot 19 \cdot 41$	10167	$3 \cdot 3389$	10207	$59 \cdot 173$
10128	$2^4 \cdot 3 \cdot 211$	10168	$2^3 \cdot 31 \cdot 41$	10208	$2^5 \cdot 11 \cdot 29$
10129	$7 \cdot 1447$	10169	10169	10209	$3 \cdot 41 \cdot 83$

Table C Continued

10210	$2 \cdot 5 \cdot 1021$	**10230**	$2 \cdot 3 \cdot 5 \cdot 11 \cdot 31$	**10250**	$2 \cdot 5^3 \cdot 41$
10211	10211	10231	$13 \cdot 787$	10251	$3^2 \cdot 17 \cdot 67$
10212	$2^2 \cdot 3 \cdot 23 \cdot 37$	10232	$2^3 \cdot 1279$	10252	$2^2 \cdot 11 \cdot 233$
10213	$7 \cdot 1459$	10233	$3^3 \cdot 379$	10253	10253
10214	$2 \cdot 5107$	10234	$2 \cdot 7 \cdot 17 \cdot 43$	10254	$2 \cdot 3 \cdot 1709$
10215	$3^2 \cdot 5 \cdot 227$	10235	$5 \cdot 23 \cdot 89$	10255	$5 \cdot 7 \cdot 293$
10216	$2^3 \cdot 1277$	10236	$2^2 \cdot 3 \cdot 853$	10256	$2^4 \cdot 641$
10217	$17 \cdot 601$	10237	$29 \cdot 353$	10257	$3 \cdot 13 \cdot 263$
10218	$2 \cdot 3 \cdot 13 \cdot 131$	10238	$2 \cdot 5119$	10258	$2 \cdot 23 \cdot 223$
10219	$11 \cdot 929$	10239	$3 \cdot 3413$	10259	10259
10220	$2^2 \cdot 5 \cdot 7 \cdot 73$	**10240**	$2^{11} \cdot 5$	**10260**	$2^2 \cdot 3^3 \cdot 5 \cdot 19$
10221	$3 \cdot 3407$	10241	$7^2 \cdot 11 \cdot 19$	10261	$31 \cdot 331$
10222	$2 \cdot 19 \cdot 269$	10242	$2 \cdot 3^2 \cdot 569$	10262	$2 \cdot 7 \cdot 733$
10223	10223	10243	10243	10263	$3 \cdot 11 \cdot 311$
10224	$2^4 \cdot 3^2 \cdot 71$	10244	$2^2 \cdot 13 \cdot 197$	10264	$2^3 \cdot 1283$
10225	$5^2 \cdot 409$	10245	$3 \cdot 5 \cdot 683$	10265	$5 \cdot 2053$
10226	$2 \cdot 5113$	10246	$2 \cdot 47 \cdot 109$	10266	$2 \cdot 3 \cdot 29 \cdot 59$
10227	$3 \cdot 7 \cdot 487$	10247	10247	10267	10267
10228	$2^2 \cdot 2557$	10248	$2^3 \cdot 3 \cdot 7 \cdot 61$	10268	$2^2 \cdot 17 \cdot 151$
10229	$53 \cdot 193$	10249	$37 \cdot 277$	10269	$3^2 \cdot 7 \cdot 163$
100000	$2^5 \cdot 5^5$	**100020**	$2^2 \cdot 3 \cdot 5 \cdot 1667$	**100040**	$2^3 \cdot 5 \cdot 41 \cdot 61$
100001	$11 \cdot 9091$	100021	$29 \cdot 3449$	100041	$3 \cdot 33347$
100002	$2 \cdot 3 \cdot 7 \cdot 2381$	100022	$2 \cdot 13 \cdot 3847$	100042	$2 \cdot 50021$
100003	100003	100023	$3 \cdot 7 \cdot 11 \cdot 433$	100043	100043
100004	$2^2 \cdot 23 \cdot 1087$	100024	$2^3 \cdot 12503$	100044	$2^2 \cdot 3^2 \cdot 7 \cdot 397$
100005	$3 \cdot 5 \cdot 59 \cdot 113$	100025	$5^2 \cdot 4001$	100045	$5 \cdot 11 \cdot 17 \cdot 107$
100006	$2 \cdot 31 \cdot 1613$	100026	$2 \cdot 3^2 \cdot 5557$	100046	$2 \cdot 50023$
100007	$97 \cdot 1031$	100027	$23 \cdot 4349$	100047	$3 \cdot 33349$
100008	$2^3 \cdot 3^3 \cdot 463$	100028	$2^2 \cdot 17 \cdot 1471$	100048	$2^4 \cdot 13^2 \cdot 37$
100009	$7^2 \cdot 13 \cdot 157$	100029	$3 \cdot 33343$	100049	100049
100010	$2 \cdot 5 \cdot 73 \cdot 137$	**100030**	$2 \cdot 5 \cdot 7 \cdot 1429$	**100050**	$2 \cdot 3 \cdot 5^2 \cdot 23 \cdot 29$
100011	$3 \cdot 17 \cdot 37 \cdot 53$	100031	$67 \cdot 1493$	100051	$7 \cdot 14293$
100012	$2^2 \cdot 11 \cdot 2273$	100032	$2^6 \cdot 3 \cdot 521$	100052	$2^2 \cdot 25013$
100013	$103 \cdot 971$	100033	$167 \cdot 599$	100053	$3^2 \cdot 11117$
100014	$2 \cdot 3 \cdot 79 \cdot 211$	100034	$2 \cdot 11 \cdot 4547$	100054	$2 \cdot 19 \cdot 2633$
100015	$5 \cdot 83 \cdot 241$	100035	$3^4 \cdot 5 \cdot 13 \cdot 19$	100055	$5 \cdot 20011$
100016	$2^4 \cdot 7 \cdot 19 \cdot 47$	100036	$2^2 \cdot 89 \cdot 281$	100056	$2^3 \cdot 3 \cdot 11 \cdot 379$
100017	$3^2 \cdot 11113$	100037	$7 \cdot 31 \cdot 461$	100057	100057
100018	$2 \cdot 43 \cdot 1163$	100038	$2 \cdot 3 \cdot 16673$	100058	$2 \cdot 7^2 \cdot 1021$
100019	100019	100039	$71 \cdot 1409$	100059	$3 \cdot 33353$

Table C

100060	$2^2 \cdot 5 \cdot 5003$	**100090**	$2 \cdot 5 \cdot 10009$	**100120**	$2^3 \cdot 5 \cdot 2503$
100061	$13 \cdot 43 \cdot 179$	100091	$101 \cdot 991$	100121	$7 \cdot 14303$
100062	$2 \cdot 3^3 \cdot 17 \cdot 109$	100092	$2^2 \cdot 3 \cdot 19 \cdot 439$	100122	$2 \cdot 3 \cdot 11 \cdot 37 \cdot 41$
100063	$47 \cdot 2129$	100093	$7 \cdot 79 \cdot 181$	100123	$59 \cdot 1697$
100064	$2^5 \cdot 53 \cdot 59$	100094	$2 \cdot 50047$	100124	$2^2 \cdot 25031$
100065	$3 \cdot 5 \cdot 7 \cdot 953$	100095	$3 \cdot 5 \cdot 6673$	100125	$3^2 \cdot 5^3 \cdot 89$
100066	$2 \cdot 50033$	100096	$2^8 \cdot 17 \cdot 23$	100126	$2 \cdot 13 \cdot 3851$
100067	$11^2 \cdot 827$	100097	$199 \cdot 503$	100127	$223 \cdot 449$
100068	$2^2 \cdot 3 \cdot 31 \cdot 269$	100098	$2 \cdot 3^2 \cdot 67 \cdot 83$	100128	$2^5 \cdot 3 \cdot 7 \cdot 149$
100069	100069	100099	$31 \cdot 3229$	100129	100129
100070	$2 \cdot 5 \cdot 10007$	**100100**	$2^2 \cdot 5^2 \cdot 7 \cdot 11 \cdot 13$	**100130**	$2 \cdot 5 \cdot 17 \cdot 19 \cdot 31$
100071	$3^2 \cdot 11119$	100101	$3 \cdot 61 \cdot 547$	100131	$3 \cdot 33377$
100072	$2^3 \cdot 7 \cdot 1787$	100102	$2 \cdot 50051$	100132	$2^2 \cdot 25033$
100073	$19 \cdot 23 \cdot 229$	100103	100103	100133	$11 \cdot 9103$
100074	$2 \cdot 3 \cdot 13 \cdot 1283$	100104	$2^3 \cdot 3 \cdot 43 \cdot 97$	100134	$2 \cdot 3^2 \cdot 5563$
100075	$5^2 \cdot 4003$	100105	$5 \cdot 20021$	100135	$5 \cdot 7 \cdot 2861$
100076	$2^2 \cdot 127 \cdot 197$	100106	$2 \cdot 50053$	100136	$2^3 \cdot 12517$
100077	$3 \cdot 33359$	100107	$3^2 \cdot 7^2 \cdot 227$	100137	$3 \cdot 29 \cdot 1151$
100078	$2 \cdot 11 \cdot 4549$	100108	$2^2 \cdot 29 \cdot 863$	100138	$2 \cdot 50069$
100079	$7 \cdot 17 \cdot 29^2$	100109	100109	100139	$13 \cdot 7703$
100080	$2^4 \cdot 3^2 \cdot 5 \cdot 139$	**100110**	$2 \cdot 3 \cdot 5 \cdot 47 \cdot 71$	**100140**	$2^2 \cdot 3 \cdot 5 \cdot 1669$
100081	$41 \cdot 2441$	100111	$11 \cdot 19 \cdot 479$	100141	$239 \cdot 419$
100082	$2 \cdot 163 \cdot 307$	100112	$2^4 \cdot 6257$	100142	$2 \cdot 7 \cdot 23 \cdot 311$
100083	$3 \cdot 73 \cdot 457$	100113	$3 \cdot 13 \cdot 17 \cdot 151$	100143	$3^3 \cdot 3709$
100084	$2^2 \cdot 131 \cdot 191$	100114	$2 \cdot 7 \cdot 7151$	100144	$2^4 \cdot 11 \cdot 569$
100085	$5 \cdot 37 \cdot 541$	100115	$5 \cdot 20023$	100145	$5 \cdot 20029$
100086	$2 \cdot 3 \cdot 7 \cdot 2383$	100116	$2^2 \cdot 3^5 \cdot 103$	100146	$2 \cdot 3 \cdot 16691$
100087	$13 \cdot 7699$	100117	$53 \cdot 1889$	100147	$17 \cdot 43 \cdot 137$
100088	$2^3 \cdot 12511$	100118	$2 \cdot 113 \cdot 443$	100148	$2^2 \cdot 25037$
100089	$3^3 \cdot 11 \cdot 337$	100119	$3 \cdot 23 \cdot 1451$	100149	$3 \cdot 7 \cdot 19 \cdot 251$

Answers to Exercises

Section 1

1. All of them.

4. 2, 5, 2.

5. 1, n.

6. d.

7. 3, 3; 3, 0.

9. 13, 34.

10. $x = 5$, $y = -6$.

Section 2

1. One, one.

3. $72 = 2^3 \cdot 3^2$, $480 = 2^5 \cdot 3 \cdot 5$.

4. $p | a_n$ for some n.

5. Suppose that the result is true for $k = r - 1$. Then $p|(a_1 a_2 \cdots a_{r-1})a_r$ implies $p|a_1 a_2 \cdots a_{r-1}$ or $p|a_r$. In the first case, the induction assumption says that $p|a_j$ for some j, $1 \le j \le r - 1$. In the second case, $p|a_r$. In either case, $p|a_j$ for some j, $1 \le j \le r$, so the result is true for $k = r$. Since the result is true for $k = 1$, it holds for all k.

6. 25, 45, 65, 81, and 85.

7. $2 \cdot 3 \cdot 5^2 \cdot 53$.

Section 3

1. The left-hand side is even; the right-hand side is odd.

2. All solutions are $x = 5t$, $y = 2 - t$, t an integer.

3. (c).

4. $x = 10 + 3t$, $y = -t$, t an integer.

5. $x = 6$, $y = 1$ and $x = 3$, $y = 2$.

Section 4

1. True, true, false, true.

2. Then $km = a - b$, so $m|(a - b)$ and $a \equiv b \pmod{m}$.

3. 1, 7, 9, 4, and 8.

4. $n \equiv 1 \pmod 2$. $n = 1 + 2k$ for some k. n leaves a remainder of 1 when divided by 2.

9. For example, $5 \cdot 4 \equiv 5 \cdot 6 \pmod{10}$, but $4 \not\equiv 6 \pmod{10}$.

10. (a) $x \equiv 2 \pmod 7$, (b) $x \equiv 4 \pmod 7$.

11. $x \equiv 2 \pmod 3$.

12. Both are wrong.

Section 5

1. For example, $4x \equiv 3$, $5x \equiv 4$, $6x \equiv 6 \pmod{12}$ have, respectively, 0, 1, and 6 solutions.

2. (b), (c), and (d) have no solutions.

3. This is included in Theorem 1.

4. (a) 2, (b) $x = -1 + 10t$, $y = 2 - 9t$.

5. 3, 1, 5, 0, 1.

6. $x = 2, 5, 8, 11$, or 14.

7. 2, 7, 12; 2; none; 2.

9. Any $x \equiv 104 \pmod{105}$.

Section 6

2. 10, 1.

3. (2, 6), (3, 4), (5, 9), (7, 8).

Section 7

1.

n	11	12	13	14	15	16
$d(n)$	2	6	2	4	4	5.

2. $d(p^3) = 4$. $d(p^n) = n + 1$.

3. $d(p^3 q) = 8$. $d(p^n q) = 2(n + 1)$.

4. 20.

5.

n	9	10	11	12	13	14
$\sigma(n)$	13	18	12	28	14	24.

6. $\sigma(p^3) = 1 + p + p^2 + p^3$. $\sigma(pq) = 1 + p + q + pq$.

8. $\sigma(p^n) = 1 + p + p^2 + \cdots + p^n = (p^{n+1} - 1)/(p - 1)$.

9. $\sigma(240) = 744$.

10.

n	13	14	15	16	17	18	19	20	21	22	23	24
$f(n)$	1	1	1	32	1	6	1	4	1	1	1	12.

Section 9

5. 1, 3. 1, 3, 5, 7. 1, 3, 5, 7, 9, 11, 13, 15. $\phi(2^n) = 2^{n-1}$, the number of odd positive integers less than 2^n.

7. $\phi(m)$.

8. 24, 36, 36.

9. (a) 12, 13, 14, 15, 16. (b) 2^k. (c) p^k.

10. $C_1 = \{1, 3, 5, 9, 11, 13\}$, $C_2 = \{2, 4, 6, 8, 10, 12\}$, $C_7 = \{7\}$, $C_{14} = \{14\}$.

Section 10

1. 2, 2, and 2.

2. 1, 2, 3, or 6. 2 and 5 have order six, 4 and 7 have order three, 8 has order two, and 1 has order one.

3. 191 (39, 77, 115, and 153 are composite).

5. 3 and 7 are primitive roots of 10.

6. The orders are 2, 4, and 1.

8. 8, 12, 15, 16, 20, 21, and 24.

9. $\text{ind}_5 1 = 0$, $\text{ind}_5 2 = 4$, $\text{ind}_5 3 = 5$, $\text{ind}_5 4 = 2$, $\text{ind}_5 5 = 1$, and $\text{ind}_5 6 = 3$.

Section 11

1. $x^2 + 4x + 3 \equiv 0 \pmod{5}$.

2. $(x + 2)^2 \equiv 1 \pmod{5}$.

3. 2 and 4.

4. 2 and $p - 2$.

5. 1, 3, 4, 5, and 9.

6. 15 and 16.

7. 1, 1, 1, 1.

8. 1, 1, 1.

9. If $p \nmid a$, then $(a^2/p) = 1$.

13. 1, 1.

14. 5, 13, 17.

15. $-1, -1$.

Section 12

2. No solution.

7. None is an odd prime.

Section 13

1. $31 = 2^4 + 2^3 + 2^2 + 2^1 + 2^0$. $33 = 2^5 + 2^0$.

2. 6, 7.

3. $n' < 2^r$. Hence $r > e_i$ for all i.

4. 9, 7, 64.

5. 10_2, 10100_2, 11001000_2.

Section 14

1. 11, 19, 110, 6χ.

3. 28, 40, 6χ6.

4. 8.

5. 371, 275.

Section 15

1. 73/4950.

2. $.\overline{17073}$.

3. 4, 2, 2.

4. 7, 6, 5.

5. $.\overline{02439}$.

6. 5.

Section 16

1. If p divides any two of x, y, and z, then it divides the third.

2. Because 2 would divide (a, b).

Section 17

1. $c^2 \equiv 2 \pmod 4$ is impossible.

6. Because $n = 2v^2$. Because $b^2 = m^2 - n^2$.

7. If $n = 0$, then $a^2 = 2mn = 0$, and a, b, c would be a trivial solution.

Section 18

2. $325 = 18^2 + 1^2$.

4. If $r = s = 0$, then $k|x$ and $k|y$. Thus $k|p$, so $k = 1$ or p. We assumed that $k > 1$ at the start. If $k = p$, then one of x and y is zero, which is a contradiction.

Section 19

1. $3 \cdot 17 = 7^2 + 1^2 + 1^2 + 0^2 = 5^2 + 4^2 + 3^2 + 1^2$.

Section 20

1. The smallest solutions are $3^2 - 2 \cdot 2^2 = 1$ and $2^2 - 3 \cdot 1^2 = 1$.

2. If $x - my = x + my = 1$, then $2x = 2$. The other case is similar.

3. 26, 15.

4. $(r + sN^{1/2})^k(r - sN^{1/2})^k = (r^2 - Ns^2)^k = 1^k = 1$.

5. Let $a = c = x_1$ and $b = d = y_1$ in Lemma 1.

Section 21

1. If $(a, b) = d$, then $d|(an + b)$ for all n, and the sequence could contain at most one prime, namely d itself, if it is prime.

3. Check cases (mod 2), (mod 3), and (mod 7).

5. Check cases or note that $n^2 + 21n + 1 \equiv (n + 1)^2$ (mod 19).

Section 22

1. 4, 5, 2, 2.

4. $\psi(32) = \log(2^4 \cdot 3^3 \cdot 5^2 \cdot 7 \cdot 11 \cdot 13 \cdot 17 \cdot 19 \cdot 23 \cdot 29) = 28.5 \ldots$

5. In the mth column, there are $[x/2^m]$ non-zero terms.

6. $\theta((x/2)^{1/m})$, where $m = [\log(x/2)/\log 2]$. $\theta((x/m)^{1/2})$, where $m = [x/2^2]$.

7. Yes; the last term is $[x/p^m]$, where $m = [\log x/\log p]$.

9. $[\log n/\log p]$.

10. 110.

13. $n = [\log x/\log 2]$.

Appendix A

1. $f(n) = (n - 1)^2$.

2. $f(n) = n^2 + 1$.

3. $f(n) = n^2 - 6n + 7$.

4. All $n \geq 17$.

5. "$1 = 1 \cdot 2/2$." Yes.

Appendix B

1. (a) $f(1)g(6) + f(2)g(5) + f(3)g(4) + f(4)g(3) + f(5)g(2) + f(6)g(1)$.
(b) $\phi(1) + \phi(2)/2 + \phi(3)/3$.
(c) $f(2)g(-1) + f(3)g(-1) + f(4)g(-1) + f(2)g(0) + f(3)g(0)$
$+ f(4)g(0) + f(2)g(1) + f(3)g(1) + f(4)g(1)$.

2. (a) $\displaystyle\sum_{i=1}^{k} i$, (b) $\displaystyle\sum_{j=1}^{17} j/(2j + 1)$, (c) $\displaystyle\sum_{k=2}^{17} k/(k - 1)^2$.

In all cases, other correct answers are possible.

3. (a) 2. (b) 2. (c) 24.

4. (a) 41. (b) 45. (c) 9801.

5. 1. (b) 2^{17}. (c) 3/14.

6. $[x + n] = [x] + n$ for any integer n.

8. 179.

Appendix C

2. 13 or 68.

3. 13 or 122.

4. 23, 41, 87, or 105.

6. 23, 41, and 55.

7. None, since $x^2 \equiv 10 \pmod{16}$ is impossible.

Hints for Problems

Section 1

2. If $a = k_1 b$ and $b = k_2 a$, then $k_1 k_2 = 1$, so $k_1 = k_2 = 1$ or $k_1 = k_2 = -1$.

4. Use Problem 3.

6. (b) If $d|n$ and $d|(n + 2)$, then $d|2$.

7. If $d|n_i$ and $d|N$, then $d|(N - n_1 n_2 \cdots n_k)$.

8. (a) If $d > 0$, $d|c$, and $d|b$, then $d|a$ and $d|b$.

10. (b) If $d|k$ and $d|(n + k)$, then $d|n$. Or, apply Lemma 4.

11. (b) If $299r + 247s = 13$, then $299(r + 247) + 247(s - 299) = 13$.

12. (b) If r/s is the root, then $(r/s)^2 + a(r/s) + b = 0$ or $r^2 + ars + bs^2 = 0$. Show that $s|r^2$ and $(r, s) = 1$ imply $s|1$ by using Theorem 4.

15. $(c/d, a/d) = 1$ and $(c/d)|(a/d)b$; apply Theorem 5.

16. Show that $d|2a$ and $d|2b$. Then apply Theorem 6.

19. Either use induction or show that of any three consecutive integers, one is divisible by 2 and one is divisible by 3.

20. (a) If 3^m has last digit 9, so does $3^m(81)^m$.

Section 2

1. (f) $111|111,111$.
   (g) Note that $10^{12} - 1 = (10^2 - 1)(10^2 + 1)(10^4 - 10^2 + 1)(10^4 + 10^2 + 1)$
   $= 99 \cdot 101 \cdot 9901 \cdot 10101$. Use Tables A and C.

4. (b) One way is to let $n = 20 + 77k$, $k = 0, 1, \ldots$. Then $7|(6n - 1)$ and $11|(6n + 1)$ for all n.

9. One way is to use Problems 6 and 8 to show that n and $n + 1$ must both be squares.

11. Write $17p = n^2 - 1 = (n + 1)(n - 1)$, and consider the cases $17 = 1, n - 1$, $n + 1$, or $n^2 - 1$.

13. Verify that
$$2^{ab} - 1 = (2^a - 1)(2^{(b-1)a} + 2^{(b-2)a} + \cdots + 1).$$

15. $n/p < n^{2/3}$. If n/p is composite, it has a prime factor less than $(n^{2/3})^{1/2} = n^{1/3}$, which is a contradiction.

19. (a) $N + ((N - 1)/2)^2 = ((N + 1)/2)^2$.

20. Suppose that n is composite. Then it has a prime divisor among $p_1, p_2, \ldots, p_k$.

22. If $p|a_i$ and $p|a_j$ $(i \neq j)$, then $p|(a_i - a_j)$, so $p|(i - j)P_n$. $p|P_n$ is impossible because $p|a_i$. $p|(i - j)$ is impossible because $-n \leq i - j \leq n$, and if $p|a_i$, then $p \geq P_n > n$.

Section 3

5. A scorpion has 8 legs, and let us assume that a centipede has 100.

Section 4

8. Cast out nines.

10. Every integer is congruent to one of $0^2, 1^2, 2^2, \ldots, 9^2$ (mod 10).

13. $(n + 1)^3 - n^3 = 3n(n + 1) + 1$.

14. Consider $3n^2 + 3n + 1$ with $n \equiv 0, 1, 2, 3, 4$ (mod 5).

16. Show that $d_k 10^k + d_{k-1}10^{k-1} + \cdots + d_0 \equiv d_k + d_{k-1} + \cdots + d_0$ (mod 3).

18. It is enough to show that $a_i \equiv a_j$ (mod p) implies $i = j$.

19. There are $3 \cdot 365 + 366 = 1461$ days between one leap year February 1st and the next. $1461 \equiv 5$ (mod 7). Remember that 2000 will be a leap year.

20. The number whose digits are *abba* is congruent to $a - b + b - a$ (mod 11).

23. Show that any cube is congruent to one of 0, 1, and 8 (mod 9), and that no combination of three of these can sum to anything congruent to 4 (mod 9).

24. Write $x^m - 1 = (x - 1)(x^{m-1} + x^{m-2} + \ldots + 1)$, and show that $m^k|(x - 1)$ and $m|(x^{m-1} + x^{m-2} + \cdots + 1)$.

25. $7 \cdot 11 \cdot 13 = 1001$. Show that $n \equiv f(n)$ (mod 1001) and hence if 7, 11, or 13 divides $f(n)$, then it divides n too.

Section 5

1. (e) From Table A, $6191 = 41 \cdot 151$, so the congruence is equivalent to $40x \equiv 191$ (mod 41) and $40x \equiv 191$ (mod 151). Thus $x \equiv 14$ (mod 41) and $x \equiv 1$ (mod 151).

4. (d) The system is the same as

$$x \equiv 3 \text{ (mod 5)}, \qquad x \equiv 3 \text{ (mod 7)}, \qquad x \equiv 3 \text{ (mod 11)}.$$

5. Add multiples of 2401 to 4 until you get a multiple of 9.

7. Solve $n \equiv 0$ (mod 3), $n \equiv 3$ (mod 5), $n \equiv 3$ (mod 7).

8. Let the number be $2^a 3^b 5^c$, and get conditions on a, b, and c.

9. Note that $x \equiv 1$ (mod 6) implies $x \equiv 1$ (mod 2) and $x \equiv 1$ (mod 3), so these two conditions are redundant.

10. (b) $n = 4k_1$ and $n = 16k_2 - 2$. It is impossible that $4k_1 \equiv 16k_2 - 2$ (mod 4).

11. We know that

$$(m + 1)x = r(m + 1) + k_1 m(m + 1),$$
$$mx = sm + k_2 m(m + 1)$$

for some k_1 and k_2. Subtract.

13. Solve

$$3a = 20r + r,$$
$$5b = 20(r + 1) + (r + 1),$$
$$7c = 20(r + 2) + (r + 2).$$

17. Apply Theorem 1.

Section 6

2. 163 is prime.

3. $7^4 \equiv 1 \pmod{10}$.

4. $7^4 \equiv 1 \pmod{100}$ too.

7. $-1 \equiv (p - 1)! \equiv (p - k)!(p - (k - 1))(p - (k - 2)) \cdots (p - 1) \pmod{p}$, and $p - r \equiv -r \pmod{p}$.

9. $(p - 1)! = (p - 1)(p - 2)(p - 3)!$.

10. Use Problem 5.

11. (a) Applying the binomial formula (see Appendix B), we have

$$(k + 1)^p = k^p + \binom{p}{1} k^{p-1} + \binom{P}{2} k^{p-2} + \cdots + \binom{p}{p-1} k + 1,$$

and $p \left| \binom{p}{r} \right.$ for $r = 1, 2, \ldots, p - 1$.

(*b*) Either use induction or add

$$1^p - 0^p \equiv 1 \pmod{p}$$
$$2^p - 1^p \equiv 1 \pmod{p}$$
$$\cdots$$
$$a^p - (a - 1)^p \equiv 1 \pmod{p}.$$

12. Fermat's Theorem says that $a^n \equiv b^n \equiv 1 \pmod{n + 1}$.

13. (b) $1 + 2 + \cdots + (p - 1) = p(p - 1)/2$.

(c) The least residues of $2, 4, 6, \ldots, 2(p - 1) \pmod{p}$ are a permutation of $1, 2, \ldots, p - 1$. Hence

$$1^m + 2^m + \cdots + (p - 1)^m \equiv 2^m + 4^m + \cdots + (2(p - 1))^m \pmod{p}$$

or

$$\sum_{i=1}^{p-1} i^m \equiv 2^m \sum_{i=1}^{p-1} i^m \pmod{p};$$

since $2^m \not\equiv 1 \pmod{p}$, we have $p \left| \sum_{i=1}^{p-1} i^m \right.$.

14. For Wilson's Theorem, set $a = 1$. To get Fermat's Theorem, we then have
$$a^p(-1) \equiv a(-1) \pmod{p}.$$

17. Show that $11^{341} - 11 \equiv 13 \pmod{31}$.

18. (a) $a^{p+q} - a^{p+1} - a^{q+1} + a^2 = (a^p - a)(a^q - a)$.
(b) $a^{pq} - a^p \equiv a^q - a \pmod{p}$, and $a^{pq} - a^q \equiv a^p - a \pmod{q}$.

19. $(2^{p-1})^2 \equiv 1 \pmod{p}$.

20. Use Fermat's Theorem and Lemma 1.

21. $1 + n + n^2 + \cdots + n^{p-2} = (n^{p-1} - 1)/(n - 1)$.

22. Apply Problem 21 with $n = 10$. Or, if you have read Section 15, note that since $p \neq 2$ or 5, $1/p$ has a repeating decimal expansion:
$$\frac{1}{p} = .\overline{d_1 d_2 \cdots d_k}.$$
It follows that
$$(10^k - 1) = p(d_1 d_2 \cdots d_k),$$
so $p|(10^k - 1)/9$.

23. Write $a = c^n + kp$. Then
$$a^{(p-1)/n} = (c^n + kp)^{(p-1)/n} \equiv c^{p-1} \pmod{p}.$$

Section 7

13. $\sum_{d|n} d = \sum_{d|n} n/d$: exactly the same terms appear in both sums, but their order is reversed.

15. Use Problem 14 and Theorem 2.

18. Put $x = 6 + a$ and $y = 6 + b$. Then $ab = 36$.

19. Put $x = N + a$ and $y = N + b$. Then $ab = N^2$. There are $d(N^2)$ positive values of a that satisfy this and $d(N^2)$ negative values, one of which is $-N$.

21. Take n to be a prime p, $p > 3$.

22. Put $x + y = a$ and $x - y = b$. Then $ab = N$.

23. Take the equation modulo 4 and consider these cases: x, y both odd; one odd and one even; and both even.

24. Mimic the text.

Section 8

6. Use Problem 5 and the fact that
$$\sum_{d|N} 1/d = \frac{1}{N} \sum_{d|N} N/d = \frac{1}{N} \sum_{d|N} d$$
for any positive integer N.

8. If p, n are amicable, then

$$p + 1 = \sigma(p) = p + n,$$

which is impossible.

9. If p^e, n are amicable, then

$$\frac{p^{e+1} - 1}{p - 1} = \sigma(p^e) = p^e + n,$$

and this implies $n = (p^e - 1)/(p - 1)$.

10. (a) $1 + p$ can have as divisors no more than $1, 2, 3, \ldots, 1 + p$, and the sum of these is $(p^2 + 3p + 2)/2$.

(b) If p^2, n are amicable, then from Problem 9, $\sigma(1 + p) = 1 + p + p^2$.

12. (f) Show that $s(pq) = 2 - (p - 1)(q - 1)$.

(g) If n is composite, then $\sigma(n) > n + 1$, so

$$\sigma(2^k(2^{k+1} - 1)) = \sigma(2^k)\sigma(2^{k+1} - 1) > (2^{k+1} - 1)2^{k+1}.$$

(h) Let $2^{k+1} - 1 = p$, a prime. Then $d = 2^n$ or $2^n p$ for some n, $0 \le n < k$. Show that $s(d) = -1$ or $2^{k+1} - 2^{n+1}$.

(i) If $p \ne 2$, show that $s(2^k p) = (2^{k+1} - 1) - p > 0$. Consider the case $p = 2$ also.

13. Show that $s(p^e) = (2p^e - p^{e+1} - 1)/(p - 1)$, and show that $2p^e \le p^{e+1} - 1$.

14. Consider these cases: $p = 2$, $p = 5$, and $p \equiv 1, 3, 7$, or $9 \pmod{10}$.

15. Show that because $2^6 \equiv 1 \pmod 9$ every prime p, $p \ge 5$, satisfies $2^{p-1} \equiv 1$ or $7 \pmod 9$. Consider $p = 3$ separately.

16. $2^{p-1}(2^p - 1) = (1/2)(4^p - 2^p) = (1/2)((3 + 1)^p - (3 - 1)^p)$; expand, using the binomial formula.

17. If $P_2 = p_2^{2e_2+1}$, then $Q_2 = \sigma(P_2)$ would be even. If $P_1 = p^{4a+3}$, then

$$Q_1 = 1 + p + p^2 + \cdots + p^{4a+3} \equiv 0 \pmod 4.$$

Section 9

8. There is less here than meets the eye: the problem is an exercise in notation.

10. Multiply both sides of the congruence by a, and apply Theorem 1.

13. $(dt, dm) = d$ if and only if $(t, m) = 1$.

14. Use the Corollary to Theorem 3: if m and n have a common factor, then

$$\prod_{p|n}\left(1 - \frac{1}{p}\right)\prod_{p|m}\left(1 - \frac{1}{p}\right)$$

has more factors, each smaller than 1, than

$$\prod_{p|mn}\left(1 - \frac{1}{p}\right).$$

16. Write $m = 2^r M$ and $n = 2^s N$, where M and N are odd. Then $(M, N) = 1$, so

$$\phi(mn) = \phi(2^{r+s}MN) = 2^{r+s-1}\phi(M)\phi(N).$$

Then calculate $\phi(m)\phi(n)$.

17. Write $m = p^r M$ and $n = p^s N$ with $(p, M) = (p, N) = 1$. Then $(M, N) = 1$; use the fact that ϕ is multiplicative, as in Problem 16.

18. Write $n = 2^k N$ with N odd. Then

$$\phi(n) = \phi(2^k)\phi(N) = 2^{k-1}\phi(N),$$

while $n/2 = 2^{k-1}N$.

19. Write $n = 2^k 3^j N$ with $(2, N) = (3, N) = 1$, and proceed as in Problem 18.

20. Write n as in Problem 19. Then

$$\phi(n) = 2^k 3^{j-1}\phi(N) \le 2^k 3^{j-1}N.$$

21. Prove that if $n - 1$ and $n + 1$ are both primes and $n > 4$, then $6|n$. Then apply Problem 20.

24. n cannot have more than one odd prime factor.

25. Show that n cannot have more than two prime factors that are distinct and that $p^{a-1}(p - 1)q^{b-1}(q - 1) = 14$ is impossible for primes p, q and positive integers a, b.

27. Show that if $1 \le m \le n$ and $(m, n) = 1$, then $(n - m, n) = 1$. Thus if $n > 2$, the integers less than n and prime to it can be paired so that the sum of each pair is n.

Section 10

7. $\phi(10021) = 9100 = 2^2 \cdot 5^2 \cdot 7 \cdot 13$.

9. (a) Apply Lemma 2, (b) 2 is a primitive root of 37.

10. If $x^2 \equiv a \pmod p$, then $2\,\mathrm{ind}_g x \equiv \mathrm{ind}_g a \pmod{p - 1}$, and $\mathrm{ind}_g a = 2\,\mathrm{ind}_g x$ or $2\,\mathrm{ind}_g x - (p - 1)$.

11. $g^{(p-1)/2} \equiv 1 \pmod p$ is impossible since g is a primitive root of p.

14. Multiply the left-hand side of the congruence by $a - 1$.

15. (a) If a is even, then $(h^a)^{(p-1)/2} \equiv 1 \pmod p$.
  (b) Let k be any primitive root of p. From Problem 15(a), $g \equiv k^a$ and $h \equiv k^b$ $\pmod p$ with a and b odd. Thus $a + b$ is even, and

$$(gh)^{(p-1)/2} \equiv (k^{(a+b)/2})^{(p-1)} \equiv 1 \pmod p,$$

so gh does not have order $p - 1 \pmod p$.

16. $(a + 1)^6 \equiv 21(a^2 + a + 1) + 1 \equiv 1 \pmod p$ because $a^2 + a + 1 \equiv 0 \pmod p$. Further, $(a + 1)^3 \equiv -1 \pmod p$, and $(a + 1)^2 \equiv a \pmod p$.

17. From Theorem 2, any prime factor must be of the form $34k + 1$. Moreover, the smallest prime factor must be less than $(131071)^{1/2}$ and so less than 362. The only primes to test are 103, 137, 239, and 307.

18. Adapt the proof of Theorem 2. a has order 1, 2, p, or $2p \pmod q$. Show that it is impossible for a to have order p. Show that if a has order 2, then $q \,|\, (a + 1)$ and if a has order $2p$, then $q \equiv 1 \pmod{2p}$.

19. From Problem 18, the primes, other than 3, that can divide $2^{19} + 1$ are of the form $38k + 1$. Since $((2^{19} + 1)/3)^{1/2} < 2^9$, the only primes to test are 191, 229, 419, and 457.

20. If $g^{n+1} \equiv g^n + d \equiv g^{n-1} + 2d \pmod{p}$, then

$$g^n(g - 1) \equiv g^{n-1}(g - 1) \pmod{p}.$$

This is impossible.

21. (a) Let $n_1, n_2, \ldots, n_{\phi(m)}$ be the positive integers less than or equal to m and relatively prime to it. Then the numbers $\mathrm{ind}_g\, n_k$, $k = 1, 2, \ldots, \phi(m)$ are a permutation of the numbers $0, 1, \ldots, \phi(m) - 1$. Hence

$$n_1 n_2 \cdots n_{\phi(m)} \equiv g^{1+2+\cdots+\phi(m)-1}$$
$$\equiv g^{\phi(m)(\phi(m)-1)/2} \equiv g^{\phi(m)/2} \equiv -1 \pmod{m}.$$

22. (b) If $\mathrm{ind}_g\, h = a$ and $\mathrm{ind}_h\, g = b$, then $g \equiv h^b \equiv (g^a)^b \pmod{m}$.

23. Apply Problem 22.

Section 11

6. They will be just the least residues of $1^2, 2^2, \ldots, 15^2$.

7. (e) $22 \equiv -1 \pmod{23}$.

12. Apply the quadratic reciprocity theorem.

17. $7 \mid (n^2 + 1)$ if and only if -1 is a quadratic residue $\pmod 7$.

18. $(1/p) = (ab/p) = (a/p)(b/p)$.

19. $159 = 3 \cdot 53$ is not prime. Consider $x^2 \equiv 211 \pmod 3$ and $x^2 \equiv 211 \pmod{53}$.

21. $(p/q) = ((q + 4a)/p) = (4a/q) = (a/q)$ and $(q/p) = ((p - 4a)/p) = (-4a/p) = (-1/p)(a/p)$. Thus $(p/q)(q/p) = (-1/p)(a/p)(a/q)$. Apply the quadratic reciprocity theorem: since $p \equiv q \pmod 4$, there are only two cases, namely $p \equiv q \equiv 1 \pmod 4$ and $p \equiv q \equiv 3 \pmod 4$.

Section 12

1. If $p > 3$, then those integers k such that the least residue $\pmod p$ of $3k$ is greater than $(p - 1)/2$ are just those k such that $(p - 1)/6 < k \le (p - 1)/3$. Consider cases: $p = 12n + 1, 5, 7,$ or 11.

2. $p = 4^n + 1 \equiv 1 \pmod 4$, so $(3/p) = (p/3) = (2/3)$. Or, note that $4^n \equiv 4 \pmod{12}$ for all positive n and apply Problem 1.

3. Show that if $p = 2^q - 1$, where q is an odd prime, then $p = 2 \cdot 4^{(q-1)/2} - 1 \equiv -1 \pmod 4$ so $(3/p) = -(p/3)$.

4. (a) 2 is a quadratic residue $\pmod p$, so $1 = (2/p) \equiv 2^{(p-1)/2} \pmod p$, by Euler's criterion.

5. (a) Note that $q \equiv 1 \pmod 4$ always.
   (b) Consider two cases: $p \equiv 1 \pmod 4$ and $p \equiv 3 \pmod 4$.

7. $((p - a)/p) = (-a/p)$.

8. The sum of the residues is congruent (mod p) to

$$1^2 + 2^2 + \cdots + \left(\frac{p-1}{2}\right)^2.$$

9. (a) $(-3/p) = (-1/p)(3/p)$; apply Problem 1.

(b) Show that if $x^2 + xr + r^2 \equiv 0 \pmod p$ then $(x + s)^2 \equiv -3s^2 \pmod p$, where s is the unique solution of $2s \equiv r \pmod p$.

10. Let n' be the unique solution of $nn' \equiv 1 \pmod p$, $n = 1, 2, \ldots, p - 1$. Then

$$n(1 + n) \equiv n^2(1 + n') \pmod p,$$

so

$$(n(n + 1)/p) = ((1 + n')/p).$$

Show that as n runs through $1, 2, \ldots, p - 2$, $1 + n'$ runs through $2, 3, \ldots, p - 1$. Then the sum we want to evaluate is

$$(2/p) + (3/p) + \cdots + ((p - 1)/p).$$

But half of $1, 2, \ldots, p - 1$ are residues and half are nonresidues, so

$$(1/p) + (2/p) + \cdots + ((p - 1)/p) = 0.$$

11. Use the theorem of Problem 7, Section 6 with $k = (p + 1)/2$.

Section 13

7. (b) One method is to use the formula for the sum of a geometric series.

8. Consider cases. b is congruent to one of 0, 1, 2, 3, and 4 modulo 5.

9. (b) $7^k \equiv 1 \pmod 2$, $k = 0, 1, 2, \ldots$.

12. See what the numbers in each list have in common when they are written in the base 2.

13. To prove that there is such a representation, induction is as good a method as any. Choose r such that

$$\frac{3^r + 1}{2} \le n < \frac{3^{r+1} + 1}{2}.$$

Then $(-3^r + 1)/2 \le n - 3^r < (3^r + 1)/2$.

14. Write 100000 in the base 2.

16. Apply Problem 13.

Section 14

16. Take $2^{1/2} = (1.4142\ldots)_x$, and convert to base 10.

17. $10^k \equiv 1 \pmod{\varepsilon}$, $k = 0, 1, \ldots$.

19. A cubic foot of water weighs 62.5_x ordinary pounds, and an ordinary gallon of water occupies 231_x cubic inches.

Section 15

1. (e) $10^2 \equiv -1 \pmod{101}$.

5. Suppose that $1/n = d_1/b + \cdots + d_t/b^t$. Then b^t/n is an integer.

6. Show that there is an integer t such that b^t/n is an integer.

8. Mimic the proof of Theorem 4.

13. (e)
$$b(rb - (r - 1)) = r(b - 1)^2 + ((r + 1)b - r),$$
$r = 0, 1, 2, \ldots$.

14. If the decimal expansion of a number is neither repeating nor terminating, then the number is not rational. Thus $.101001000100001 \ldots$ is irrational in any base, 7 in particular.

Section 16

7. If $n + (n + 1) = m^2$, then $n^2 + m^2 = (n + 1)^2$.

8. If $a^2 + b^2 = c^2$ and $ab = 2c$, then show that $(a + b)^2 = (c + 2)^2 - 4 = c(c + 4)$. $(c, c + 4) = 1, 2,$ or 4; apply Lemma 2.

9. If $m \equiv 0 \pmod 5$ or $n \equiv 0 \pmod 5$, then $5 \mid a$. If $m \equiv \pm n \pmod 5$, then $5 \mid b$. Show that in the remaining cases, $5 \mid c$.

10. Show that $4 \mid 2mn(m^2 - n^2)$ and $3 \mid 2mn(m^2 - n^2)$.

11. Combine Problems 9 and 10.

12. (a) The quadrilateral has two right angles.

13. Let $n = t(t - 1)/2$ and $m = t(t + 1)/2$ and calculate $m^2 - n^2$.

15. If $(a - d)^2 + a^2 = (a + d)^2$, then $a(a - 4d) = 0$.

16. (b) If $a^2 + b^2 = (b + 1)^2$, then $a^2 = 2b + 1$. Hence a is odd, say $a = 2n + 1$.

17. (b) We want nontrivial solutions of $m_1 n_1(m_1^2 - n_1^2) = m_2 n_2(m_2^2 - n_2^2)$. These are not easy to find.

18. Consider the equation modulo 2.

19. $2n^2 + 2n + 1 \equiv 0 \pmod k$ only when $4n^2 + 4n + 1 \equiv -1 \pmod k$.

21. If $9 = (a/c)^2 + (b/c)^2$, then $a^2 + b^2 = 9c^2$. Hence $a^2 + b^2 \equiv 0 \pmod 9$; show that this implies $a = 3r$ and $b = 3s$, and thus $r^2 + s^2 = c^2$, which has infinitely many solutions.

22. (a) If a is even, find m and n such that $a = 2mn$. If a is odd, find m and n such that $a = m^2 - n^2 = (m + n)(m - n)$.

23. If $a^2 + b^2 = c^2$ and $2(a + b + c) = ab$, show that $b = 4 + 8/(a - 4)$, so that $a = 5, 6, 8,$ or 12.

Section 17

3. Remember Fermat's Theorem.

4. If $p \nmid xyz$, then Fermat's Theorem says that
$$x^{p-1} \equiv y^{p-1} \equiv z^{p-1} \equiv 1 \ (\text{mod } p).$$

7. (b) Show that x, y, and z must all be even.

10. Since $x = y$ is impossible, assume that $x > y$. Show that $x^n < z^n$ and
$$(x + 1)^n \geq x^n + nx^{n-1} > x^n + y^n = z^n.$$
Thus $x^n < z^n < (x + 1)^n$, which is impossible.

Section 18

3. Use Table C or B.

4. $x^2 \equiv 0, 1, 2,$ or $4 \ (\text{mod } 7)$, so $x^2 + y^2 \equiv 0 \ (\text{mod } 7)$ only if $x \equiv y \equiv 0 \ (\text{mod } 7)$. But then $49 \mid (x^2 + y^2)$.

5. $x^2 \equiv 0, 1, 4,$ or $7 \ (\text{mod } 9)$.

7. Consider the prime-power decomposition of n/m.

12. $x^2 \equiv 0, 1,$ or $4 \ (\text{mod } 8)$.

13. Suppose that $4^e(8k + 7) = x^2 + y^2 + z^2$. Apply Problem 11 e times to get $8k + 7 = x_1^2 + y_1^2 + z_1^2$ for integers x_1, y_1, z_1. Then apply Problem 12.

15. $(r^2 + ws^2)(x^2 + wy^2) = (rx + wsy)^2 + w(ry - sx)^2$.

Section 19

3. $5,725,841 = 11^2 \cdot 47321$.

6. Consider three cases: at least three of x, y, z, w divisible by 3, just two divisible by 3, and at least three not divisible by 3.

11. If $k_1, k_2, \ldots, k_r$ are odd, then $k_1^2 + k_2^2 + \cdots + k_r^2 \equiv r \ (\text{mod } 8)$.

12. Consider the equation modulo 4.

Section 20

3. $x^2 + 2xy - 2y^2 = (x + y)^2 - 3y^2$.

5. Complete the square.

7. (a) The area of a triangle with sides a, b, and c is
$$(s(s - a)(s - b)(s - c))^{1/2},$$
where $s = (a + b + c)/2$.

(b) If $3a^2 - 3 = b^2$, then $3 \mid b$. Hence $b = 3c$, so $3a^2 - 3 = 9c^2$ or $a^2 - 3c^2 = 1$, a Fermat equation.

(d) Consider $3(2a + 1)^2 - 4 = c^2$ modulo 4.

9. $x_1 + y_1N^{1/2} > 1$ and $1 = x_1^2 - Ny_1^2 = (x_1 + y_1N^{1/2})(x_1 - y_1N^{1/2})$.

13. Apply Lemma 2.

14. If $(x - 1)^2 + x^2 + (x + 1)^2 = u^2 + (u + 1)^2$, then $(2u + 1)^2 = 6x^2 + 3$, so $2u + 1 = 3y$ and $3y^2 - 2x^2 = 1$. Trial will disclose solutions larger than $x = 11, y = 9$.

15. $1 + n + n^2 = m^2$ implies $4m^2 - (2n + 1)^2 = 3$. Factor the left-hand side. Or, note that $n^2 < n^2 + n + 1 < (n + 1)^2$.

Section 21

2. Note that it is not required that $f(n) \neq f(m)$ if $n \neq m$.

3. One method is to consider all possible cases for n (mod 7) and (mod 11).

5. Which primes can have -2 as a quadratic residue?

6. $n^2 + 2n + 3 = (n + 1)^2 + 2$.

7. Consider cases, modulo 2, 3, and 5.

8. Complete the square in $n^2 + n + 41 \equiv 0 \pmod{p}$.

9. Apply the binomial formula.

11. Use Wilson's Theorem.

13. Show that the condition is $n(n + 1) \mid 2n!$, and consider these cases: $n + 1$ even, $n + 1$ an odd square, and $n + 1$ odd and composite. Show that in each case, the factors of $n(n + 1)$ appear among those of $2n!$.

14. If $x + (x + 1) + \cdots + (x + t) = p$, then $(t + 1)(2x + t) = 2p$. Show that this implies $t = 1$.

15. (d) Show that n occurs $\phi(n)$ times in the nth line. This is not easy.

16. Use logarithms.

Section 23

15. Show that $3 \mid (p^2 + 2)$.

23. Consider $ab(a + b)(a - b)$ modulo 3 and modulo 8.

28. $(2/7)^2 + (5/7) = (2/7) + (5/7)^2$.

33. (a) If $n = a_0 + a_1 \cdot 12 + \cdots + a_k \cdot 12^k$, then

$$m = a_k + a_{k-1} \cdot 12 + \cdots + a_0 \cdot 12^k;$$

calculate $n - m$ (mod 11).

34. (a) If m is composite, then one of $p = 2, 3, 5, 7$ divides m. But then $p \mid (210n + m)$ too.

35. (a) If $3p + 1 = a^2$, then $3p = (a + 1)(a - 1)$.
    (b) If $3p + 2 = a^2$, then $a^2 \equiv 2 \pmod 3$.

36. Note that there is no loss of generality in assuming that $(a, b) = 1$. Complete the square on the right-hand side and see what happens (mod 4).

37. The well-known Rational Root Theorem says that if r/s is a root of
$$a_n x^n + a_{n-1} x^{n-1} + \cdots + a_1 x + a_0 = 0,$$
then $r \mid a_0$ and $s \mid a_n$. So, if r/s is a root of
$$a^2 + b^2 = 2(a + b)x + x^2,$$
then $r \mid (a^2 + b^2)$ and $s \mid 1$. Note that if x^2 is rational and $a + b \neq 0$, then
$$x = \frac{a^2 + b^2 - x^2}{2(a + b)}$$
is rational.

38. Consider $n - 1$ modulo m, 4, and 9.

39. 20413 has the prime-power decomposition $137 \cdot 149$.

40. (a) If $n = ab$, then $10^n - 1$ is composite.

43. Consider $2^{p-1}(2^p - 1) \pmod{12}$.

46. If all else fails, use induction.

48. Show that the first of a pair of twin primes must be congruent to 5 (mod 6).

51. Consider cases (mod 24).

52. $9^{10} \equiv 1 \pmod{100}$.

53. If $kp + b = m^2$, then $m^2 \equiv b \equiv c^2 \pmod{p}$, so $m \equiv \pm c \pmod{p}$. That is, for some integer n, $m = \pm c + np$.

54. $3 \mid (2^m + 1)$.

55. $10^{3n+1} - 1 = (10^{3n} - 1)(10^{2 \cdot 3n} + 10^{3n} + 1)$.

57. $z^6 - x^6 = (z - x)(z^2 + xz + x^2)(z^3 + x^3)$. Show that at least one of the factors on the right is odd.

58. (a) Complete the square.

59. (c)
$$ab\left(\frac{a + b}{2}\right)\left(\frac{a - b}{2}\right) = \left(\frac{c}{2}\right)^2.$$

60. If $n(n + 1)/2 = x^2$, then $(2n + 1)^2 = 8x^2 + 1$. If $y = 2n + 1$, then $y^2 - 8x^2 = 1$, a Fermat equation with generator $3 + 8^{1/2}$ and infinitely many solutions.

61. Take x so that $n - x^2 = m$ is odd and positive. Then let $y = (m + 1)/2$.

64. Show that if one, two, or all three of x, y, and z are odd, then the left-hand side of the equation is odd.

65. Set $m = 2pM_1$, $n = 2qN_1$, $m + 1 = 3qM_2$, and $n + 1 = 3pN_2$, where p and q are distinct primes.

66. (a) Consider both sides (mod 3).

(b) $n^2 + (n + 1)^2 + \cdots + (n + k)^2 = (k + 1)n^2 + k(k + 1)n + \sum_{r=1}^{k} r^2$.

67. For each 5 that appears as a factor of $n!$, there are two even factors.

68. $n + (n + 1) + \cdots + (n + d) = (2n + d)(d + 1)/2$; if this is a power of 2, show that d can be neither odd nor even.

70. Consider the number (mod 10).

71. Let the integers be $a - 4b, a - 3b, \ldots, a + 4b$. Then their sum of squares is $9a^2 + 60b^2$.

73. $\phi(n) \le n\left(1 - \dfrac{1}{p}\right)$, where p is the smallest prime that divides n, and $p \le n^{1/2}$.

76. If p is prime, then from Wilson's Theorem,

$$4((p - 1)! + 1) \equiv -p(p + 1)p((p - 1)! + 1)$$
$$\equiv -p((p + 1)! + 2) \quad (\bmod\, p + 2),$$

and $(p + 1)! \equiv -1 \pmod{p + 2}$ if and only if $p + 2$ is prime.

81. If $p > 12$, then $p - 9$ is even and composite.

82. If $p^{a-1}(p - 1) = q^{b-1}(q - 1)$ and $a > 1$, then $p \mid (q - 1)$, which contradicts $p > q$.

85. (a) Put $y = x + d$, $z = x + 2d$. Then $(d + x)(3d - 2x) = 0$.
(b) If $y = x + d$ and $z = x + 2d$, then $(d + x)(3d - (k + 1)x) = 0$.

87. The harmonic mean m of $a_1, a_2, \ldots, a_k$ is given by

$$\frac{1}{m} = \frac{1}{k}\left(\frac{1}{a_1} + \frac{1}{a_2} + \cdots + \frac{1}{a_k}\right).$$

Hence, for the divisors of an even perfect number n,

$$\frac{1}{m} = \frac{1}{d(n)}\sum_{d\mid n}\frac{1}{d} = \frac{1}{nd(n)}\sum_{d\mid n}\frac{n}{d} = \frac{2}{d(n)}.$$

88. (b) If $n \equiv 1 \pmod 3$, then $3 \mid p^r$, so $p = 3$. But $n \equiv 1, 4,$ or $7 \pmod 9$, so $n^2 + n + 1 \equiv 3 \pmod 9$. Hence $r = n = 1$.
(c) If $r = 2k$, then $(2n + 1)^2 - (2p^k)^2 = 3$. Factor the left-hand side.
(d) Suppose that $p \ne 3$ and $p \equiv 2 \pmod 3$. Since r is odd, $p^r \equiv 2 \pmod 3$. Moreover, $n^2 + n + 1 \equiv 1 \pmod 3$.

89. If $f(r/s) = 0$, then $r \mid a_n$ and $s \mid a_0$. Hence r and s are odd. But this implies that $0 = s^n f(r/s)$ is the sum of some even integers and an odd number of odd integers, which is impossible.

90. Find r and s such that $(p - 1)r + 1 = sn$. Then put $x = n^s$.

91. (a) If $x^n \equiv a$, then $1 \equiv x^{p-1} \equiv (x^n)^{(p-1)/n} \equiv a^{(p-1)/n} \pmod p$.

92. Induction, using the identity

$$2^{3^{k+1}} + 1 = (2^{3^k} + 1)(2^{2\cdot 3^k} - 2^{3^k} + 1),$$

is one method that will work.

93. (a) Suppose not. Then $2n^2 + 2n + (1 - 3m^2) = 0$, so

$$2n = -1 \pm (6m^2 - 1)^{1/2}.$$

Hence $6m^2 - 1 = r^2$ for some r. But $r^2 \equiv -1 \pmod 6$ is impossible.

94. Let $m = p_1 p_2 \cdots p_k$ with $p_1 < p_2 < \cdots < p_k$. Then $p_2 \mid m$ implies $(p_2 - 1) \mid m$, so $p_2 - 1 = p_1$. Thus $p_1 = 2$ and $p_2 = 3$. Further, $p_3 \mid m$ implies $(p_3 - 1) \mid m$, so $(p_3 - 1) \mid p_1 p_2$, whence $p_3 = 7$. Similarly, $(p_4 - 1) \mid 42$, so $p_4 = 43$. Finally, $(p_5 - 1) \mid 2 \cdot 3 \cdot 7 \cdot 43$, but there is no such prime.

96. The primes in the product divide $(2n)!$ but not $n!$.

98. $n^{2^m} \equiv n^{p-1} \equiv 1 \pmod{p}$, and

$$-1 = (n/p) \equiv n^{(p-1)/2} \equiv n^{2^{m-1}} \pmod{p},$$

so n has order $2^m \pmod{p}$.

103. (a) $x^2 = 10^n rs + x$.

104. Suppose that n is prime. From Wilson's Theorem, $1 + (n - 1)! = kn$ for some k. Hence

$$\alpha^{j(1+(n-1)!)} = \alpha^{jkn} = (\alpha^n)^{jk} = 1$$

for all j. If n is composite and greater than 4, then $1 + (n - 1)! = 1 + kn$ for some k, so

$$\alpha^{j(1+(n-1)!)} = \alpha^j,$$

and the sum is a geometric progression that adds to zero. The sum is also zero if $n = 4$.

105. Apply Fermat's Theorem: there is a progression starting at any term with ratio 2^{p-1}.

Appendix A

4. $1^3 + 2^3 + \cdots + n^3 = (n(n + 1)/2)^2$ for $n = 1, 2, \ldots$.

8. $(n - 1)n(n + 1)(n + 2) = (n^2 + n - 1)^2 - 1$ is an identity.

10. $(n + 1)(n + 2)(n + 3) = n(n + 1)(n + 2) + 3(n + 1)(n + 2)$, and $(n + 1)(n + 2)$ is even.

Appendix B

19. Count how many times each prime appears on either side.

20. Apply Problem 15:

$$\frac{1}{n} \sum_{d \mid n} d = \sum_{d \mid n} \frac{1}{d} \quad \text{and} \quad \frac{1}{n^2} \sum_{d \mid n} d^2 = \sum_{d \mid n} 1/d^2.$$

23. Multiply both sides of the equation by $n_1 n_2 \cdots n_k$.

Answers to Problems

Section 1

1. (a) 1. (b) 1. (c) 592. (d) 73.

5. For example, $2 \mid (-4)$.

6. (b) 1 or 2. (c) A positive divisor of k.

9. (a) $x = -40, y = 79$. (b) $x = 37, y = -73$. (c) $x = 2, y = -1$.
(d) $x = -10, y = 1$. In all cases there are also other solutions.

10. (b) Yes. (c) Yes.

11. (b) $x = 5 + 19t$, $y = -6 - 23t$, where t is an integer, give all solutions.
(c) $x = 1 + 19t$, $y = -1 - 23t$, where t is an integer, give all solutions.

17. For example, $4 \nmid 6$, but $(4, 6) = 2$.

20. (b) $m = 2, 6, 10, 14, \ldots$.

Section 2

1. (a) $3 \cdot 37$. (b) $2 \cdot 617$. (c) $5 \cdot 7 \cdot 67$. (d) $2^7 \cdot 3^3$.
(e) Prime. (f) $3 \cdot 7 \cdot 11 \cdot 13 \cdot 37$. (g) $3^3 \cdot 7 \cdot 11 \cdot 13 \cdot 37 \cdot 101 \cdot 9901$.

2. For example, $6 \mid 4 \cdot 9$, but $6 \nmid 4$ and $6 \nmid 9$.

3. $511 = 7 \cdot 73$.

4. (a) The first example occurs at $n = 20$: $119 = 7 \cdot 17$ and $121 = 11^2$. There are others at $n = 24, 31,$ and 36.

5. No.

7. n is a kth power if and only if each exponent in its prime-power decomposition is divisible by k.

8. No.

9. The result follows from $n^2 < n(n + 1) < (n + 1)^2$ for all $n > 0$.

10. (b) No, because $78 \leq \dfrac{25ab}{32} \leq 82$, no matter what the digits a and b are.

11. $p = 19$ only.

12. (a) 7. (c) If $n = 1001! + 1$, then $n + 1, n + 2, \ldots, n + 1000$ are composite.

14. No: $2^{11} - 1 = 2047 = 23 \cdot 89$ is composite (use Table A), and 11 is not.

16. False: $3 \mid 60$, $5 \mid 60$, and both 3 and 5 are greater than $60^{1/4} = 2.78 \ldots$, but $60/3 \cdot 5 = 4$ is not prime. The statement would be true if p and q were the least prime factors of n.

17. (a) 60, $2p^2q$.

19. (b) $b + a$ and $b - a$. (c) $1189 = 29 \cdot 41$. (d) $9379 = 83 \cdot 113$.

Section 3

1. (a) $x = 1 + t, y = 1 - t$. (c) $x = -1 + 16t, y = 2 - 15t$.
   (b) $x = 1 + t, y = -2t$. (d) No solutions.

In (a), (b), and (c), t is any integer. There are other ways of writing the solutions. In (c) for example, $x = -17 + 16s, y = 17 - 15s$, s an integer, gives the same set of solutions.

2. (a) $x, y = 1, 1$. (c) $x, y = 1, 3$ or $6, 1$.
   (b) No solutions. (d) $x, y = 3, 2$.

3. (a) $x = -4 - 5t, y = -5 - 2t, t = 0, 1, 2, \ldots$.
   (b) No solutions.

4. $x, y, z = 22, 8, 1; 23, 6, 2; 24, 4, 3;$ or $25, 2, 4$.

5. 21 worms.

6. Nine apples at 9 cents each and three oranges at 6 cents each.

7. 5 cows.

8. Four ways. There may be 14, 15, 16, or 17 quarters.

9. 8 sophomores, 15 juniors, 3 seniors.

10. A has \$10, B has \$75, and C has \$15.

11. Ann is 5 and Mary 12, unless both are zero.

13. $x = -1 - 9m + 8n, y = 1 + 9m - 7n, z = -m$, where m and n are arbitrary integers is one way of writing the solutions. Another is

$$x = 7 + 63m + 8n, \qquad y = -6 - 54m - 7n, \qquad z = -m.$$

14. \$25.51.

15. The lower price was 3 cents per egg, and the smallest yield was \$1.20. The largest was \$2.00.

Section 4

2. $2^2 \equiv 3^2 \pmod 5$, but $2 \not\equiv 3 \pmod 5$ is one counterexample.

3. False: $1 \equiv 4 \pmod 3$, but $1^2 \not\equiv 4^2 \pmod 9$ is one counterexample.

4. 3.

7. 1, 7, 11, 13, 17, 19, 23, or 29.

8. 6.

16. (c) If the sum of the digits of an integer, taken in reverse order with alternating signs, is divisible by 11, then the integer is divisible by 11. Another way of stating this is as follows. Let a be the sum of the digits in the even-numbered places of n. Let b be the sum of the other digits. If $11 \mid (a - b)$, then $11 \mid n$.

17. A.

19. 1996, 2024, 2052, and 2080.

20. Every palindrome with an even number of digits is divisible by 11.

22. Any $n \equiv 15 \pmod{30}$ satisfies the three congruences.

24. Since $x \equiv 1 \pmod{m}$.
$$x^{m-1} + x^{m-2} + \cdots + 1 \equiv 1 + 1 + \cdots + 1 \ (m \text{ terms})$$
$$\equiv 0 \pmod{m}.$$

25. $118{,}050{,}660 = 2^2 \cdot 3^2 \cdot 5 \cdot 7 \cdot 13 \cdot 7207$.

Section 5

1. (a) 9. (b) 6. (c) 2, 8, or 14. (d) 6 or 15. (e) 6041.

2. For example, $15x \equiv 14$, $13x \equiv 14$, $12x \equiv 14$, and $20x \equiv 0 \pmod{20}$ have, respectively, 0, 1, 4, and 20 solutions.

3. 0, 1, 2, 4, 5, 10, or 20 solutions.

4. (a) $x \equiv 1 \pmod 6$.   (d) $x \equiv 3 \pmod{385}$.
   (b) $x \equiv 13 \pmod{42}$.   (e) $x \equiv 605 \pmod{1066}$.
   (c) $x \equiv 348 \pmod{385}$.

5. $x \equiv 534 \pmod{2401}$.

6. 17, 157, or other larger and more unlikely numbers.

7. 213.

8. $2^{15}3^{10}5^6 = 30{,}233{,}088{,}000{,}000$ is one answer. Zero is another, but not as good.

9. 301 or any number congruent to it $\pmod{420}$.

10. (a) 2223 or any number congruent to it $\pmod{3600}$. (b) No.

12. (a) $x = 3$, $y = 0$. (b) $x = 5$, $y = 5$.

13. 28, 21, 18; 63, 42, 33; and 98, 63, 48.

15. 62.

16. $(m_i, m_j) \mid (a_i - a_j)$ for all i, j. It can be shown that this condition is also sufficient for the system to have a solution.

17. (a, b).

Section 6

1. 6.

2. 1.

3. 3.

4. 43.

7. $(p - k)!(k - 1)! \equiv (-1)^k \pmod p$, $k = 0, 1, \ldots, p$.

8. (a) 2, 0, 0, 0, 0. (b) If $n > 4$ is composite, then $(n - 1)! \equiv 0 \pmod{n}$.

10. (b) For example, $p = 5, r = 3$ or $p = 7, r = 3$ or 5.

20. 1 or -1.

21. All n such that $n \not\equiv 0$ or 1 $\pmod{p}$.

Section 7

1. (a) 8. (b) 24. (c) 48.

2. (a) 96. (b) 1344. (c) 14880.

3. (a) 4. (b) 24. (c) 4.

4. (a) $74 \cdot 138 = 10{,}212$. (b) $12 \cdot 9092 = 109{,}104$.
(c) $15 \cdot 13 \cdot 140 = 27{,}300$.

8. For example, $d(n) - n$ is not multiplicative.

9. 24, 48.

10. Yes: $d(2^{k-1}) = k$.

11. $2^4 \cdot 3^2 \cdot 5 \cdot 7 = 5040$ is one. Others are $2^4 \cdot 3^2 \cdot 5 \cdot 11 = 7920$ and $2^4 \cdot 3^2 \cdot 5 \cdot 13 = 9360$.

12. $d(p^{59}) = 60$ for any prime p.

14. Even k.

15. All n of the form $2^e p_1^{2e_1} p_2^{2e_2} \cdots p_k^{2e_k}$.

18. Sixteen solutions are given by $(x, y) = (2, -3), (3, -6), (4, -12), (5, -30), (7, 42), (8, 24), (9, 18), (10, 15)$ and their reversals; the seventeenth is $(x, y) = (12, 12)$.

19. $2d(N^2) - 1$.

22. $2d(N)$.

24. If $n = p_1^{e_1} p_2^{e_2} \cdots p_k^{e_k}$, then $\sigma_2(n) = (1 + p_1^2 + p_1^4 + \cdots + p_1^{2e_1})(1 + p_2^2 + \cdots + p_2^{2e_2}) \cdots (1 + p_k^2 + \cdots + p_k^{2e_k})$.

25. If

$$n = \prod_{p|n} p^{e_p},$$

then

$$\sigma_k(n) = \prod_{p|n} \frac{p^{(e_p + 1)k} - 1}{p^k - 1}.$$

Section 8

3. $2^{p-1}(2^p - 1) = n(n + 1)/2$ with $n = 2^p - 1$.

4. Yes. $2^{p-1}(2^p - 1) = 1 + 2 + \cdots + (2^p - 1)$.

12. (b) 6 is perfect; 12, 18, and 20 are abundant; the rest are deficient.
(c) $s(945) = 30$.

(d)

n	3	4	5	6	7	8	9	10
$n(n+1)$	12	20	30	42	56	72	90	110
$\sigma(n(n+1))$	28	42	72	96	120	195	234	216
$s(n(n+1))$	4	2	12	10	8	51	54	−4.

Section 9

1. 12, 96, 960.

2. 9792, 90900.

4. 1, 5, 7, 11, 13, 17.

5. None: $\phi(n) \leq n$ for all n.

7. The sum of the positive integers less than n and relatively prime to it is $n\phi(n)/2$.

11. (a) $p - 2$. (b) $p^2 - 2p$. (c) Let $\psi(n)$ be the number of elements in the sequence $1 \cdot 2, 2 \cdot 3, \ldots, n(n+1)$ which are relatively prime to n. If $n = p^k$, p an odd prime, then $\psi(n) = p^{k-1}(p - 2)$. ψ is multiplicative; thus $\psi(n)$ can be found for any positive integer n.

12. All solutions are 17, 32, 34, 40, 48, and 60.

15. (a) 1, 1, 2, 1, 2, 2, 2,   (c) k,
    (b) $k - 1$,   (d) $j + k - 1$

17. $\phi(mn) = (p/(p - 1))\phi(m)\phi(n)$.

23. (a) 0, −13, 0, −15, 0,
    (b) $-p$,
    (c) 0,   (d) $-p^k$,
    (e) $\displaystyle\sum_{d|n} (-1)^{n/d}\phi(d) = \begin{cases} -n & \text{if } n \text{ is odd} \\ 0 & \text{if } n \text{ is even.} \end{cases}$

24. $n = 1, 2, 4, p^k$, or $2p^k$, where $p \equiv 3 \pmod 4$ and k is a positive integer.

26. All such $2k$, $k < 50$, are 14, 26, 34, 38, 50, 62, 68, 74, 76, 86, 90, 94, and 98.

28. (a) No: at least one corner will have two even coordinates.
    (b) The one nearest the origin has corners (14, 20), (14, 21), (15, 20), and (15, 21).
    (c) $2\phi(p + 1) - 3$.

Section 10

1. (a) 199, 202.   (b) 10198, 10201, 10202.

2. 4, 2, 4, 4, 2, 4, and 2. (Hence 15 has no primitive roots.)

3. (a)

n	1	2	3	4	5	6	7	8	9	10	11	12	13	14
ind$_2 n$	0	1	5	2	22	6	12	3	10	23	25	7	18	13
n	15	16	17	18	19	20	21	22	23	24	25	26	27	28
ind$_2 n$	27	4	21	11	9	24	17	26	20	8	16	19	15	14

(b) 26.   (c) 11.

4. (a) 1. (b) 0. (c) 3.

5. (c) No.

6. No.

7. Yes.

9. (b) The primitive roots of 37 are the least residues (mod 37) of 2^k, where $(k, 36) = 1$, namely

$$2, 2^5, 2^7, 2^{11}, 2^{13}, 2^{17}, 2^{19}, 2^{23}, 2^{25}, 2^{29}, 2^{31}, \text{ and } 2^{35}.$$

The primitive roots of 37 are thus

$$2, 5, 13, 15, 17, 18, 19, 20, 22, 24, 32, \text{ and } 35.$$

10. (a) 2, 6, 7, and 8. (b) 7, 3, 1, and 9, respectively.

12. (b) $p - 4$.

13. 6 and 26.

21. (b) $m = 8$ is one example.

22. (a) If g and h are primitive roots of m, then

$$(\text{ind}_g h)(\text{ind}_h g) \equiv 1 \pmod{\phi(m)}.$$

23. $g = h$.

Section 11

1. (a), (b), (c), and (d).

2. (a) 22 and 31. (b) 2 and 5. (c) 13 and 18. (d) 25 and 9948.

3. (a) No solutions. (b) 0 and 4. (c) 2.

4. (a) Four: 1, 7, 9, and 15. (b) No: 16 is not an odd prime.

5. (a) 3 and 6. (b) 4 and 5.

6. 1, 2, 4, 5, 7, 8, 9, 10, 14, 16, 18, 19, 20, 25, 28.

7. (a) 1. (b) 1. (c) 1. (d) -1. (e) -1. (f) -1.

8. 3, 11, and 17.

9. (a) -1. (b) -1. (c) -1. (d) 1.

10. (a) 1. (b) 1.

13. Yes: 21 and 76. Yes: 16 and 37.

15. A residue.

16. Yes.

17. $(-1/7) = -1$; A loses.

19. Yes: 23, 76, 83, and 136 are all solutions.

20. $(r/p) = 1$.

Section 12

1.
$$(3/p) = \begin{cases} 1 & \text{if } p \equiv 1 \text{ or } 11 \text{ (mod 12)} \\ -1 & \text{if } p \equiv 5 \text{ or } 7 \text{ (mod 12)}. \end{cases}$$

4. (b) 167.

6. (a) $p = 2$, and those odd p with $(-1/p) = 1$. That is, $p = 2$ or $p \equiv 1$ (mod 4).
(b) All: $p \mid (p^2 + p)$.   (c) Those congruent to 1 (mod 4).

7. (b) Then p and $p - a$ are both residues or nonresidues.

Section 13

1. (a) 10111010100_2.   (b) 2001021_3.   (c) 4231_7.
(d) 2037_9.   (e) 1137_{11}.

2. (a) 421.   (b) 709.   (c) 1107.   (d) 2305.

3. (a) 102.   (b) 153.   (c) 7.

4.

+	1	2	3	4	5	6	10
5	6	10	11	12	13	14	15
6	10	11	12	13	14	15	16
10	11	12	13	14	15	16	20.

5.

	2	3	4	5	6	10
5	13	21	26	34	42	50
6	15	24	33	42	51	60.

6. (Answers are base 7 numbers.) (a) 105.   (b) 1445.   (c) 534.   (d) 54421.

7. (a) 19/49.   (b) 1/2.   (c) 13/16.

9. (c) Any base with $b \equiv 1$ (mod 2).

10. (b) $121_b = (b + 1)^2$.   (c) $(b + 3)^2$.

12. For example, $19 = 16 + 2 + 1$ and 19 appears in lists 16, 2, and 1.

14. 6 gifts in all: one each of $32, $128, $512, $1024, $32768, and $65536.

15. (b) The first, second, third, fourth, and last.

16. Weights of 1, 3, 9, 27, and 81 pounds will weigh objects up to 121 pounds.

Section 14

1. $32 = 2 \cdot 17$, $33 = 3 \cdot 11$, $34 = 2^3 \cdot 5$, $36 = 2 \cdot 3 \cdot 7$, $38 = 2^2 \cdot \varepsilon$, $39 = 3^2 \cdot 5$, $3\chi = 2 \cdot 1\varepsilon$, $40 = 2^4 \cdot 3$, and 31, 35, 37, and 3ε are prime.

2. (a) $8\chi 67$.   (b) $-27\chi 5$.   (c) $15\varepsilon 43126$.   (d) $.658$; remainder $19\chi 0$.

3. (a) 2454.   (b) 156000.   (c) $1/3^3 = 1/23 = .054$.

6. $4\varepsilon.\varepsilon6 = (59.95833\ldots)_\chi$.

χ. The possible last digits of a power, greater than 1, of n are as follows:

n	0	1	2	3	4	5	6	7	8
last digits	0	1	4,8	3,9	4	1,5	0	1,7	4,8

n	9	χ	ε
last digits	9	4	ε.

ε. Except for the primes 2 and 3, only the digits 1, 5, 7, and ε.

12. (a) False, true, false.

13. No: $55 = 5 \cdot 11$.

14. $24/9$, $9/\varepsilon\varepsilon$.

15. $3.\overline{86\chi351}$, $.\overline{0\varepsilon}$.

16. $2^{1/2} = 1.4\varepsilon792\ldots$; 1.5 is a very close approximation.

18. If an integer has digits $d_k d_{k-1} \cdots d_1 d_0$, then it is divisible by 7, 11, or 17 if $(d_2 d_1 d_0) - (d_5 d_4 d_3) + (d_8 d_7 d_6) - \cdots$ is divisible by 7, 11, or 17.

19. The digits in the rest of this answer are in the base χ. One do-metric mile is $1728/1760 = .98\ldots$ ordinary miles. One cubic yard holds 1728 do-metric pints and 1616 ordinary pints; the do-metric pint is $.93\ldots$ ordinary pints. One cubic yard holds 1728 do-metric pounds and approximately $27(62.5) = 1687.5$ ordinary pounds of water; the do-metric pound is $.98\ldots$ ordinary pounds.

1χ. (a) 265; 266 in leap years.

(b) Only 260, 261, ..., 269.

(d) No: if the number is $abcd$, then in base χ we have

$$1728a + 144b + 12c + d = 1000(a + 1) + 100b + 10c + d$$

or

$$264a + 22b + c = 500,$$

and this has no solutions with $0 \le a \le 9$, $0 \le b \le 9$, and $0 \le c \le 9$.

Section 15

1. (a) 2. (b) 3. (c) 1. (d) 3. (e) 4. (f) 6.

2. (a) 2. (b) 3.

3. Yes. $31415 = 5 \cdot 6283$, so the length of the period of $1/31415$ is at most 6282. In fact, $6283 = 61 \cdot 103$, and $10^{3060} \equiv 1 \pmod{6283}$.

7. $1/16$, $1/18$, and $1/24$.

9. (a) 2. (b) 4. (c) 3. (d) 6. (e) 10.

10. (a) $.\overline{01}$. (b) $.\overline{0011}$. (c) $.\overline{000111}$.

11. (a) 6. (b) 1. (c) 16.

12. (a) $.\overline{0\varepsilon}$. (b) $.0\overline{\chi35186}$.

13. (a) $.\overline{012345679}$. (b) $.\overline{0123457}$. (c) $.\overline{012346}$.

(d) In base b, $1/(b - 1)^2 = .\overline{012 \ldots (b - 4)(b - 3)(b - 1)}$.

Section 16

1. Eleven more:

a	8	12	16	20	24	28	32	36	24	36	16
b	6	9	12	15	18	21	24	27	10	15	30
c	10	15	20	25	30	35	40	45	26	39	34.

2. One is given by $m = 10$, $n = 3$.

4. (b) $m = 9$, $n = 4$.

5. Yes.

9. Yes.

11. This follows from Problems 9 and 10.

12. (a) 234. (b) Another has sides 33, 56, 63, and 16, with one diagonal 65.
(c) Yes: paste together Pythagorean triangles with the same hypotenuse. To
see that there are infinitely many, whenever c is divisible by 5 (to get this, take
$m \equiv 1 \pmod 5$ and $n \equiv 2 \pmod 5$), say $c = 5k$, let a and b be the other sides
of the fundamental solution. Since $3k$, $4k$, $5k$ is also a Pythagorean triangle
with hypotenuse $5k$, the quadrilateral with sides a, b, $3k$, $4k$ has integer sides
and area $3abk^2$.

14. $(n - 1)^2 + n^2 = (n + 1)^2$ implies $n^2 = 4n$, so, since $n > 0$, $n = 4$.

16. (a) $(2n + 1)^2 + (2n(n + 1))^2 = (2n(n + 1) + 1)^2$, $n = 0, 1, \ldots$.

17. (a) (20, 21, 29) and (12, 35, 36) have area 210.
(b) The triangles with generators 35, 11 and 33, 23 have the same area.
(c) Let the sides of the two triangles be a, b, c and a_1, b_1, c. Then $a^2 + b^2 = a_1^2 + b_1^2$ and $ab = a_1 b_1$. These imply $a = a_1$ and $b = b_1$.

21. $9 = (12/5)^2 + (9/5)^2 = (36/13)^2 + (15/13)^2 = (24/17)^2 + (45/17)^2 = \ldots$.

22. (b) $13^2 + 84^2 = 85^2$ and $14^2 + 48^2 = 50^2$, for example.

23. 5, 12, 13 and 6, 8, 10 are the only solutions.

24. (b) $696^2 + 697^2 = 985^2$ is the next.

Section 17

5. No.

6. All even k.

8. Yes. n^2 and $n + 1$ are relatively prime.

9. $x = y = z = 2$ is one solution. $x = y = 2^{16}$, $z = 2^{13}$ is a solution derived
from Problem 8 with $a = b = 1$, $r = 4$, $s = 13$.

11. Let $x = (ac)^{rn}$, $y = (bc)^{rn}$, $z = c^s$, where

$$c = a^{rn^2} + b^{rn^2} \quad \text{and} \quad rn^2 + 1 = (n - 1)s.$$

12. As in Problem 11, but $rn^2 + 1 = ms$. This has solutions if $(n^2, m) = 1$.

Section 18

1. $153 = 12^2 + 3^2$.

2. $1970 = 41^2 + 17^2$.

3. $10045 = 98^2 + 21^2 = 91^2 + 42^2$; $10048 = 88^2 + 48^2$;
   $10049 = 100^2 + 7^2 = 95^2 + 32^2$.

6. No.

10. 102, 103, 111, 119, 124, 127, 135, 143.

14. 37, 149, . . . are missed.

15. The generalization is false: $(-5/3) = 1$, but $x^2 + 5y^2 = 3$ is impossible.

16. If $n = x(x + 1)/2 + y(y + 1)/2$, then $4n + 1 = (x + y + 1)^2 + (x - y)^2$.

17. If $n = (a/c)^2 + (b/c)^2$, then $c^2n = a^2 + b^2$, and c^2n is representable if and only if n is. The answer is, "the same integers as in Theorem 1."

Section 19

1. If $n = x^2 + y^2 + z^2 + w^2$, then we have

n	2	3	5	7	11
x, y, z, w	1, 1, 0, 0	1, 1, 1, 0	2, 1, 0, 0	2, 1, 1, 1	3, 1, 1, 0

n	13	17
x, y, z, w	3, 2, 0, 0 or 2, 2, 2, 1	4, 1, 0, 0 or 3, 2, 2, 0

n	19	23
x, y, z, w	4, 1, 1, 1 or 3, 3, 1, 0	3, 3, 2, 1.

2. (a) $121 = 11^2 + 0^2 + 0^2 + 0^2 = 10^2 + 4^2 + 2^2 + 1^2 = \cdots$
   $= 7^2 + 6^2 + 6^2 + 0^2$.
   (b) $391 = 19^2 + 5^2 + 2^2 + 1^2$.
   (c) $47321 = 217^2 + 14^2 + 6^2 + 0^2$.
   In (a), (b), and (c), there are several other representations.

3. $2387^2 + 154^2 + 66^2 + 0^2$, among others.

4. $112^2 + 63^2 + 35^2 + 21^2$.

7. 11, 14, and 15.

9. $170^2 + 68^2 + 4^2 + 1^2$.

Section 20

1. (a) $5 + 2 \cdot 6^{1/2}$. (d) $19 + 6 \cdot 10^{1/2}$.
   (b) $8 + 3 \cdot 7^{1/2}$. (e) $10 + 3 \cdot 11^{1/2}$.
   (c) $3 + 8^{1/2}$. (f) $7 + 2 \cdot 12^{1/2}$.

2. The two smallest positive nontrivial solutions are
   (a) 5, 2; 49, 20. (d) 15, 4; 449, 120.
   (b) 3, 1; 17, 6. (e) 8, 1; 127, 16.
   (c) 7, 2; 97, 28. (f) 10, 1; 199, 20.

3. The three smallest positive solutions are $x, y = 1, 1; 3, 4; 11, 15$.

4. $x_k, y_k, k = 1, 2, \ldots$, is a solution, where

$$(x_k + y_k) + y_k \cdot 3^{1/2} = (2 + 3^{1/2})^k.$$

5. (c) The equation becomes

$$x + ay = 1 \quad \text{or} \quad x + ay = -1;$$

either has infinitely many solutions.

6. (a) m_k and n_k are generators for the triangle, and they can be determined from

$$m_k - n_k + n_k \cdot 2^{1/2} = (3 + 2 \cdot 2^{1/2})^k$$

for any k.

(b) $(3, 4, 5)$, $(696, 697, 993)$.

7. (b) $(3, 4, 5)$, area 6; $(13, 14, 15)$, area 84; $(51, 52, 53)$, area 1170.

10. x_k/y_k gets closer and closer to $N^{1/2}$.

11. $x_k/y_k = 3/2, 17/12, 99/70$, and $577/408$ for $k = 1, 2, 3, 4$. The difference between x_k/y_k and $2^{1/2}$ is about .09, .002, .00007, and .000002, respectively.

14. The next solution is $108^2 + 109^2 + 110^2 = 133^2 + 134^2$.

17. $x_5 = 6(577) - 99 = 3363$, $y_5 = 6(408) - 70 = 2378$, and

$$\frac{3363}{2378} = 1.41421360 \ldots.$$

Section 21

1. $f(x) = (x^2 + x + 4)/2$.

2. $y = 3$, for example.

3. $n^2 + 21n + 1 \equiv n^2 + 1 \pmod 7$, and -1 is not a quadratic residue (mod 7). $n^2 + 21n + 1 \equiv (n + 5)^2 - 2 \pmod{11}$, and 2 is not a quadratic residue (mod 11).

5. $p \equiv 1$ or 3 (mod 8); also $p = 2$.

6. $p \equiv 1$ or 3 (mod 8); also $p = 2$.

7. 11, 17, 41, 47, 71, and 77.

16. Approximately 10^{38}.

Section 22

2.

$$\frac{1}{6} \log \log x \le \sum_{p \le x} \frac{1}{p} \le \frac{10}{3} \log \log x.$$

3.

$$\frac{1}{3} \log N \le \sum_{n \le N} (\log p_n)/p_n \le \frac{10}{3} (\log N + 1).$$

5. (b) r_p.

Section 23

2. 40 sixes, 10 fives, and 21 tens.

3. $2^{27} - 1 = 7 \cdot 73 \cdot 262657$.

5. (a) 0, 1, 4, or 7, (b) No. It is congruent to 2 (mod 9).

6. (b) If $a = 2$, $b = 3$, and $p = 5$, then $(a + b)'$ does not even exist, much less be congruent to $a' + b'$ (mod p).

7. 115, 117, 119, 121, 123, 125.

9. Pascal had discovered the identity

$$(n + 1)^3 - n^3 - 1 = 6 \left(\frac{n(n + 1)}{2} \right).$$

11. The first numbers on the left-hand sides are every other triangular number. The result may be written

$$(2n^2 + n)^2 + \cdots + (2n^2 + 2n)^2 = (2n^2 + 2n + 1)^2 + \cdots + (2n^2 + 3n)^2.$$

12. (a) 9, (b) 90, (c) If $k = 2n$, there are $9 \cdot 10^{n-1}$; if $k = 2n - 1$, there are $9 \cdot 10^{n-1}$.

13. (c) No. There are countless counterexamples.

14. For those k such that $(k, k(k + 1)/2) = 1$; namely, $k = 1$ or 2.

16. 1 man, 5 women, and 14 children.

17. (o) Not true for ordinary divisibility are (l), (m), and (n).

18. $f(n) = (3 + (-1)^{n+1})/4$ is one. $f(n) = (n + 1 - 2[n/2])/2$ is another.

19. 788; 210998.

21. (a) $533 = 13 \cdot 41$. (b) $1073 = 29 \cdot 37$.

22. $170833 = 412^2 + 33^2 = 407^2 + 72^2 = 393^2 + 128^2 = 348^2 + 223^2$
$$= 13 \cdot 17 \cdot 773.$$
$182410 = 427^2 + 9^2 = 423^2 + 59^2 = 409^2 + 123^2 = 401^2 + 147^2$
$$= 383^2 + 189^2 = 381^2 + 193^2 = 347^2 + 249^2 = 303^2 + 301^2$$
$$= 2 \cdot 5 \cdot 17 \cdot 29 \cdot 37.$$

25. Those with sides 3, 4, 5, 6, 8, 10, 12, 15, 16, 17, 20, 24, 30, 32, and 34.

26. $x^2 + (x + 1)^2 = (x(x + 1) + 1)^2 - (x(x + 1))^2$.

28. No. $x^2 + (1 - x) = x + (1 - x)^2$ for all x.

29. (a) $x = y = 2^{1/4}$, $z = 2^{1/2}$ for example.
(b) If $(a/b)^4 + (c/d)^4 = (r/s)^4$, then $(ads)^4 + (bcs)^4 = (bdr)^4$, which is impossible.

$$\left(a + \frac{b}{c} \right)^{1/2} = a \left(\frac{b}{c} \right)^{1/2}$$

if and only if $b/c = a/(a^2 - 1)$. All numbers $a + a/(a^2 - 1)$ have the same property as $5 + 5/24$.

32. No. It neither repeats nor terminates.

33. (b) $(b - 1) \mid (n - m)$.

35. (a) $p = 5$, (b) None.

37. If $a + b \neq 0$, then x^2 is irrational, and if $a = -b$, then $x^2 = a^2 + b^2$ is rational.

39. \$137, 7 years ago, at 7%.

40. No: $111 = 3 \cdot 37$.

42. None.

45. (c) $f(2^n p_1^{e_1} p_2^{e_2} \cdots p_k^{e_k}) = (e_1 + 1)(e_2 + 1) \cdots (e_k + 1)$.

50. (a) 1, 3, 5, 8, 15, 24, 40, 120; 1, 5, 8, 9, 40, 45, 72, 360.
(b) Integers that are products of distinct primes.
(c) 2^k. (d) 6, 60, and 90 are the only such numbers less than 10^{12}.

56. (c) 18, 20, and 24 are examples.

58. (b) 2, and those congruent to 1 (mod 4).
(d) $a = r^2 + s^2$, $b = r^2 - s^2$.
(e) a and b are squares, say $a = t^2$, $b = u^2$.

60. 6^2, 35^2, 204^2, 1189^2.

63. If $n \equiv 2 \pmod 4$, then $n = x^2 - y^2$ is impossible.

65. The smallest solution is $m = 14$ and $n = 20$.

69. (a)

n	2	3	4	5	6	7	8	9	10	11	12	13	14	15
$f(n)$	2	3	4	5	3	7	6	6	5	11	4	13	7	5
n	16	17	18	19	20
$f(n)$	8	17	6	19	5.

70. Its last digit is 3.

71. For example, $2^2 + 5^2 + 8^2 + \cdots + 23^2 + 26^2 = 48^2$.

72. (a) 9,876,543,210. (b) 98,763,210. (c) 4312. (d) 987,652,413.

74. $2^{d(n)}$.

75. (a) $4n^2 - 3n + 1$.   (d) $(n, -n)$.
(b) $4n^2 - 2n + 1$.   (e) $(-n + 1, n)$.
(c) $4n^2 + n + 1$.   (f) $(9, 16)$.

77. (a) $m = 4n(n + 1)$. (b) Yes.

78. False. Take $n = 4$, $m = 3$ for an example.

79. If $(a, m) = d$, the least residues (mod m) of $a, 2a, \ldots, ma$ consist of m/d numbers such that $d \mid m$, each repeated d times.

80. $P_n \equiv 3 \pmod 4$. Note that P_n is prime for $n = 1, 2, 3, 4, 5$, but $P_6 = 59 \cdot 509$. Another guess for a formula for primes may thus be discarded.

81. 1, 2, 3, 5, 7, and 11.

83. Yes. $2^{2k-2}(2^{2k-1} - 1) = 1^3 + 3^3 + \cdots + (2^k - 1)^3$ is true for all k, as may be shown using the fact that

$$1^3 + 2^3 + \cdots + n^3 = (n(n + 1)/2)^2.$$

84. If $n = d_0 + d_1 \cdot 10 + d_2 \cdot 10^2 + \cdots$, $0 \le d_i \le 9$, then $37 \mid n$ if and only if $37 \mid (d_0 d_1 d_2) + (d_3 d_4 d_5) + \cdots$.

85. (a) $x = 3t$, $y = 5t$, $z = 7t$ for any nonzero integer t.
(b) If $k \not\equiv 2$ (mod 3), then $x = 3t$, $y = (k + 4)t$, $z = (2k + 5)t$; if $k \equiv 2$ (mod 3), then $x = t$, $y = (k + 4)t/3$, $z = (2k + 5)t/3$, for nonzero integers t.

86. $x = (a + b)/2$, $y = (a - b)/2$.

88. The only solution of this equation with $r > 1$ and $n < 180,000$ is $18^2 + 18 + 1 = 7^3$.

91. (b) No: $2^6 \equiv 2$ (mod 31).

93. (b) It is necessary that -1 be a quadratic residue (mod $2k$).

94. (b) Yes: 1806. (c) No others.

95. $6. This astonishing problem rests on the fact that if the tens' digit of a square is odd, then its units' digit is 6. To observe this, see Table B. To prove it, enumerate cases.

99. Solutions are

x	10	10	14	14	17	17	21	21	31	36	44	105
y	6	35	7	34	7	34	6	35	41	45	52	111.

103. (b) $r = 78$, $s = 5$ gives $x = 625$ and $x^2 = 390625$.

Appendix A

2. $t_n = n(n + 1)/2$.

5. $1^3 + 3^3 + \cdots + (2k - 1)^3 = k^2(2k^2 - 1)$.

9. $(n + 1)(6n^2 + 9n + 2)/2$.

11. $f(n) = 17(n - 1)(n - 2)(n - 3)(n - 4)/24$ is one infinitely many such functions.

12. No. $8t_n + 1 = (2n + 1)^2$ for all n.

Appendix B

1. (a) 112. (b) 1. (c) 2. (d) 27. (e) 34. (f) 328.

2. (a) $\sum_{i=1}^{13} (2i + 1)^2$. (b) $\sum_{j=n}^{2n} 1/j$. (c) $\sum_{k=0}^{n} a_k x^k$.

(d) $\sum_{t=0}^{r} u_{r-t} v_{n-r+t}$. In all cases, other forms for the answer are possible.

3. (a) and (b) are true and may be easily proved. (c) is false: $a_k = b_k = 1$, $k = 1, 2, \ldots, n$ is one of infinitely many counterexamples.

4. (a) $3^{1/2}$. (b) 1. (c) 8640. (d) 42. (e) 46.

5. (a) 40,320. (b) 70. (c) 792.

7. (a) $x^6 + 12x^5 + 60x^4 + 160x^3 + 240x^2 + 192x + 64$.
  (b) $1 - 7a + 21a^2 - 35a^3 + 35a^4 - 21a^5 + 7a^6 - a^7$.

8. (a) $(x + 3)^3$. (b) $(x - 1)^n$.

9. 1, 2, 3, 5.

10. (a) No. (b) No.

11. $\|x\| = [x + 1/2]$.

12. 0 and -1.

13. (a) $p^2 - p$. (b) $n - [n/p]$.

14. 19.

21. $[y] - [x]$.

Appendix C

1. (a) 68, 175. (b) 138, 487. (c) 23, 105, 151, 223.

2. No solutions.

3. (a) 4. (b) 6. (c) 4.

4. (a) 33, 117, 183, 267, 333, 417, 483, 567, 633, 717, 783, 867.
  (b) 33, 217, 283, 467, 533, 717, 783, 967.
  (c) 33, 517, 583, 1067.

5. 20.

7. 0, 1, 2, 4, or 8.

8. $3^{e-1} - r$ and $3^e - r$ satisfy $x^3 \equiv -a \pmod{p}$.

9. (a) 13. (b) 4, 13, 22.

Index